Alternative Energy

Alternative Energy

Political, Economic, and Social Feasibility

Christopher A. Simon

ROWMAN & LITTLEFIELD PUBLISHERS, INC.
Lanham • Boulder • New York • Toronto • Plymouth, UK

ROWMAN & LITTLEFIELD PUBLISHERS, INC.

Published in the United States of America
by Rowman & Littlefield Publishers, Inc.
A wholly owned subsidary of The Rowman & Littlefield Publishing Group, Inc.
4501 Forbes Boulevard, Suite 200, Lanham, Maryland 20706
www.rowmanlittlefield.com

Estover Road
Plymouth PL6 7PY
United Kingdom

British Library Cataloguing in Publication Information Available

Library of Congress Cataloging-in-Publication Data

Simon, Christopher A., 1968–
　Alternative energy : political, economic, and social feasibility / Christopher A. Simon.
　　p. cm.
　Includes index.
　ISBN-13: 978-0-7425-4908-1 (cloth : alk. paper)
　ISBN-10: 0-7425-4908-9 (cloth : alk. paper)
　ISBN-13: 978-0-7425-4909-8 (pbk. : alk. paper)
　ISBN-10: 0-7425-4909-7 (pbk. : alk. paper)
　1. Energy policy—United States. 2. Renewable energy sources—United States. 3.
Power resources—Government policy—United States. I. Title.
　HD9502.U52S544 2007
　333.79′4—dc22　　　　　　　　　　　　　　　　　　　　　　　　2006026149

Printed in the United States of America

♾TM The paper used in this publication meets the minimum requirements of American
National Standard for Information Sciences—Permanence of Paper for Printed Library
Materials, ANSI/NISO Z39.48-1992.

To Mom and Dad

Contents

List of Illustrations   ix

Foreword   xi

Acknowledgments   xv

1   Why Alternative Energy and Fuels?   1

2   Studying Public Policy and Alternative Energy/Fuels   21

3   Overview of Alternative Energy and Fuels and Their Uses   39

4   Alternative Energy/Fuels as a Public Policy Innovation   63

5   Solar Energy   87

6   Wind Energy   103

7   Geothermal Energy   123

8   New Century Fuels and Their Uses   145

9   Historical Precedents: Alternative Energy/Fuels and
    Legitimacy Issues   169

10 Public Policy, Institutional Developments, and
    Policy Interests   189

11 Practical Demonstration of the E
    Alternative Energy   209

Conclusion: The Future of Alternative Energy 219

Index 227

About the Author 233

List of Illustrations

1.1	Domestic Petroleum Extraction in the United States (1958–2005)	6
2.1	Rivalry and the Market for Gasoline	23
2.2	Lowi's Policy Typology	32
3.1	Wind Turbines and Visual Pollution	49
4.1	Hydrogen Technology and Public Policy	79
5.1	Schematic of Solar Cell	89
5.2	Domestic Photovoltaic Shipments, United States	94
6.1	Wind Turbine Construction	104
6.2	Wind Turbine Blades	105
6.3	Blade Mass and Cost of Rotors	113
6.4	WPA Activity Matrix	116
7.1	Flash Steam Power Plant	124
7.2	Geothermal Use, 1975–2005	127
7.3	A Geothermal System	129
7.4	Federal Grants for Geothermal Projects, 2000–2003	136
9.1	Risk and Policy Solutions	172
11.1	HOMER® Optimum Systems	215
C.1	Social and Environmental Costs of Fossil Energy Use	222

Foreword

In October 2005, Washington State University, Spokane, and the Thomas S. Foley Institute for Public Policy and Public Service, Washington State University, Pullman, sponsored a conference titled *Global Oil Depletion and Implications for the Pacific Northwest*, which was held at the Davenport Hotel. The conference featured keynote speakers Matt Simmons, Herman Franssen, and Roger Bezdek. Simmons is the author of *Twilight in the Desert: The Coming Saudi Oil Shock and the World Economy* (2005). Franssen and Bezdek are energy economists with considerable domestic and international experience. Based on information currently known, all three individuals agreed that global production of conventional petroleum has either peaked or was likely to peak within the next few decades. As Simmons pointed out, the sustainability of the Saudi oilfields is indeterminate because Saudi Arabian oil reserve records may not be accurate and require better verification such that analysts are able to gauge the future of oil production in the world's largest oil producing nation. All three of these experts agreed that fuel supply would directly impact the energy paradigm upon which most of the world economy is based, particularly in terms of fossil fuels as stored energy. Reliance on fossil fuel as a stored energy to either operate equipment or for thermal and electrical purposes is rising despite the potential for sharp and continual declines in supply. All other things being equal, the laws of economics point to rising prices for petroleum and increases in unmet demand. To avoid rising prices, demand for fossil fuel must be curtailed and the development of viable replacement fuels must rapidly proceed. Unmet demand has major implications for global economic development and for social and economic justice.

While there is a strong sense that markets will adjust eventually, evolution of the energy market will take time. Bezdek pointed out that the world must develop alternative sources of energy supply to meet a near probable energy deficit produced by oil depletion—and solutions must be developed now if policy and technology innovations are to become feasible within a generation.

Supply and demand issues will require identification and establishment of viable energy sources and decreased demand. Thermal and electrical energy demand are relatively inelastic in the modern world, because of the nature of the technologies that are central to meeting other types of demand. There are few large scale, viable substitutes to meet demand for thermal and electrical energy in the United States. Nuclear energy, while an increasingly subject of discussion, has reached its supply limit within the regulatory boundaries established in the post-Three Mile Island policy environment.

The Spokane conference and similar conferences held in Seattle, Washington, and Chicago, Illinois, proposed many solutions to excess demand for energy, as well as supply related issues. Presenters addressed alternative energy supply solutions ranging from "green" sources to the expansion of nuclear power. Borrowing from Kingdon's agenda setting model, the problem stream is widely understood and solutions are subject to consent—the consent of actors in the marketplace, as expressed through economic choices and political and social consent of the governed. Consent, however, requires information and much of this book, therefore, discusses, the technical, economic, and political feasibility of alternative energy supply solutions.

One approach to establishing consent is driven by the sustainable communities model of future development in society. In essence, supply and demand functions may be established through collective action and choices rather than simply through prices. Political choices and social and political values, rather than purely free market economics, play a substantial role in shaping outcomes. In order to promote broad-based support for energy as a common resource, the "sustainable communities" model seeks collaboratively derived solutions at the local level.

Establishing a viable collaborative effort requires the meeting of at least two conditions:

First, it requires a strong and lasting commitment on the part of stakeholders to cultivate a process built on the idea of mutual trust and mutual benefit.

Commitment is a function of incentives, both individual and collective. In this context, incentives must result in action, such that individual and group actors have a strong sense that choices they make will result in greater benefit than cost. Incentives may emerge from public and private sources. Nobel laureate economist Douglass C. North documented the role that government and the private sector play in shaping social and economic conditions and change. Incentive and disincentive structures emerging from public policy, markets, institutional and individual choices made within the confines of various institutions, and policy environments interact to shape outcomes. Ultimately, incentives and disincentives impact choices and outcomes in the marketplace and may actually result in the creation or weakening of market sectors.

> Second, it requires a policy analysis integration that will bring policy stakeholders together rather than produce divisions largely based on stakeholder technical knowledge and expertise. In other words, the continued process of building a sustainable community requires transparency such that all stakeholders can access information and produce feasible alternative policy solutions that will be discussed on the basis of their relative merits rather than simply being the function of technology.

A Roman saying is fitting: *Scientia potestas est* (knowledge is power). When knowledge or access to the power of knowledge is not uniform, then influence and outcomes are likely to be biased. The sustainable communities model, however, requires that all stakeholders participate equally and that outcomes are a function of true collaboration-based agreement.

Two different sets of forces are afoot, working to develop a vision of the energy future. First, public and private sector experts—technologists and economists—are seeking to produce the next generation of alternative energy solutions to meet the coming energy gap between supply and demand, through both policy incentives and market-based innovation. Second, concerned citizens are beginning to more actively consider the need for sustainable communities. These two forces, however, do not easily mesh in a collaborative venue because experts often speak a different "language" that is not easily understood by citizen stakeholders.

The book is an effort to get everyone on the same page. There is a need to start thinking, speaking, and making decisions in a common language about the policy and market sectors related to alternative energy and fuels. The author does not intend the book to be a thorough analysis of technical issues, but will provide an accurate overview of the general issues related to alternative energy sources. Tying together economic and technical issues, the book highlights an accessible software tool that can be used to help citizens move beyond simple understanding and reactive policy participation—energy policy "shareware" makes it possible for the reader (the citizen stakeholder and/or the more technically-minded expert) to become a proactive decision maker—someone who understands alternatives and makes choices—in a setting that offers equal access and with a unified policy language.

Ultimately, feasibility and choice must be considered with the following points in mind:

- Energy policy operates in the context of past, present, and future conditions.
- Real alternatives exist.
- Trade-offs must be understood and considered.
- Costs must be recognized.
- Choices must be made in an informed manner, through the combined efforts of technical experts and citizen stakeholders.

Christopher A. Simon
University of Nevada, Reno

Acknowledgments

There are several people I wish to acknowledge. My parents Raffi G. and Susan M. Simon have always been there to support me through the writing process. My father is a biochemist—our discussions during manuscript development were particularly lively and extremely informative. I was fortunate to have James Murphy, a research professor in metallurgy at University of Nevada, review the manuscript. I appreciated his very helpful advice and enjoyed our conversations about alternative energy. Brent Steel, a political science professor at Oregon State University has been an influence in my professional life since my undergraduate days in Corvallis, Oregon. I was pleased to have his careful eyes trained upon the manuscript and appreciated his advice. My colleague and research collaborator Eric B. Herzik sparked my interest in alternative energy by inviting me to work with him on a hydrogen mass transit study conducted through the Transportation Technology Transfer Center (T2), Department of Civil and Environmental Engineering, University of Nevada. Peter Lilienthal, Ph.D., Senior Economist for International Programs, National Renewable Energy Laboratory and Tom Lambert, P.E., Mistaya Engineering were very helpful in terms of HOMER®[1] applications. John Lund at Geo-Heat Center, Oregon Institute of Technology provided valuable information regarding geothermal resource applications. Special thanks to the *Journal of Transport Economics and Policy* (Bath, UK) and the Program in Science and Global Security (Princeton, US). Research colleague and former advisor Nicholas P. Lovrich, Washington State University, helped me to establish contact with several key individuals and organizations in energy policy, to include the WSU Energy Program. Brian Romer, my editor at Rowman & Littlefield, provided valuable support. I appreciate the advice of the anonymous reviewers, who offered critical guidance in manuscript development. Special thanks to Karen Ackermann and Sarah Stanton, both at Rowman & Littlefield.

NOTE

1. "The HOMER® software was developed with U.S. government funds." (Authorized representative for the National Renewable Energy Laboratory: Paul J. White, Senior Patent Counsel).

Chapter One

Why Alternative Energy and Fuels?

INTRODUCTION

The twentieth century saw an unprecedented expansion in industrialization. In 1900, automobiles were not a primary means of transportation and air travel was nonexistent. Homes were heated and illuminated by wood, coal, and whale oil rather than by electricity and natural gas. Initially, industrial expansion was limited to Western countries; presently, industrial expansion is a global event. Entering the twenty-first century, Japan is the world's second leading economy. China is ranked among the top seven nations in gross domestic product (World Bank 2006). This rapid growth is built on a foundation of fossil energy. Many nations in the Middle East, Africa, and South America have economies largely dependent on the extraction of—rather than consumption of—fossil fuels.

The global expansion of energy use increases dependence on finite energy sources. Energy availability and ready access is a critical element in modern economies, but economic development is directly related to the ability to exercise a whole host of economic and social freedoms. A finite resource base may mean finiteness in terms of potential horizons for economic and social freedom and a sense that energy use and access is a zero-sum game rather than a basic building block for modern economic and social evolution.

Fossil fuels, one of the cheapest forms of stored energy, are often viewed as a commodity rather than a pure public good. A public good is a common resource necessary for individual and collective existence that cannot be easily marketed without causing uneven distribution, deprivation, and social and economic injustice. The overuse or misuse of a public good could create a tragedy of the commons—resource overuse, degradation, and depletion negatively impacting a large number of individuals (see Hardin 1968).

When fossil energy is readily available and relatively cheap, it is unlikely to become the subject of large-scale public policy efforts to determine distribution and use, which frequently occurs when a particular benefit is viewed from the public good perspective. However, when fossil energy is commonly viewed as expensive and scarce, it becomes the subject of political debate: Is it a pure public good? A marketable public good? Should its distribution and use be regulated?

Beyond issues of sustainability and just distribution of a good, usage-related impacts on the natural environment have become a secondary method of moving energy use and distribution into the public good debate. Fossil energy policy is a cross-cutting issue in terms of both access to fuel resources as well as fuel emissions impacts on another public good—namely, the ecology of the planet. Clean air and water and maintaining ecological diversity are essentially linked to the use of fossil energy. Policy, however, has focused to lesser degree on distribution and to a greater degree on regulation of energy use impacts. The secondary view of fossil energy places less emphasis on incentives for distribution and greater emphasis on disincentives, through tax and regulation policies, for its continued widespread use.

The challenges facing energy demand and supply potential, in general, and fossil energy, in particular, are numerous and could be characterized as "wicked problems" (Van Beuren et al. 2003). Wicked problems are issues that test the limits of human understanding and whose costs are not well understood by the marketplace or by government. What makes a problem wicked? First, wicked problems are characterized by high levels of information uncertainty. The validity and reliability of the information are unclear. In a systems approach, information is given meaning by the paradigm that governs it. From the wicked-problems perspective, information and its meaning are legitimized in a network setting, are highly relative to the given circumstances of network nodes, and change over time. Second, wicked problems exist in a contestable policy environment where multiple actors compete with each other. Finally, institutional relationships, information, and choices are made at multiple nodal points increasing the level of uncertainty about the nature of a "problem," information validity and meaning, and an acceptable solution. In essence, "wicked" problems are the same problems or circumstances that have continually existed (see Douglas and Wildavsky 1973), but demand growth factors (e.g., see Berk 1981), technological growth, and the rise of network-based solutions have made the policy environment more complex.

Wicked problems of the twenty-first century are wide ranging and are often a function of past human choices and the externalities or unintended consequences that emerge from such choices. Neither market nor public policy solutions are clearly defined. A good example of an externality that has become a wicked problem is the impending energy crisis that will have a transnational impact on the world's social, political, and economic condition. Experts predict that much of Earth's sweet crude petroleum reserves—the optimum petroleum for the manufacture of gasoline and other combustible fossil fuels—will be depleted by 2050. The world economy is built on the use of petroleum and other fossil fuels. One type of externality that must be considered is the impending need to retool an entire economic base to better fit other sources of energy that are more readily available, some of which are fossil fuels while others are alternative or renewable energy sources. The wicked problems of twenty-first-century energy policy will require a tremendous commitment of time and resources to develop very complex public-sector and market-based solutions, solutions that are simultaneously equitable to stakeholders as well as marketable and affordable to consumers.

Many of the energy policy challenges and marketable solutions of the new century and millennium can only be overcome through objective empiricism, which is the rational analysis of objects and phenomena as they exist in a measurable and describable state. Through empirical study, science must progress rapidly in order to establish comprehension of understudied phenomena or conditions related to energy supply. Other challenges, however, are most immediately related to our normative vision of the world and what is right and wrong, just and unjust, and what could be improved about the conditions under which we live, the nature of our society, and the condition of the natural environment. While empiricism is concerned with things as they *are*, normative issues are concerned with things as they *ought to* be, according to our individual and collective views of the "good" life and "good" society—questions that transcend both public policy and marketplace capitalism (see Bellah, et al. 1992; Etzioni 1988; Heilbroner 1994; Simon 2007).

The wicked problems of our new century, however, challenge us to think beyond the historically bifurcated world of the empirical and the normative. Empiricists have often criticized normatively driven scholars on the basis that without knowing what "is," it would be very difficult to determine how things "ought to be." Conversely, normatively driven scholars criticize empiricists for being "abstract empiricists" (see Mills 1967), constructing knowledge in a narrow utilitarian manner absent any sense of the larger meaning and implication of scientific endeavor. The development of alternative energy solutions to meet increasing demand, however, is no longer simply a function of the utility of the marketplace and the supply and demand curves of economic theory. Rather, the wicked problems facing energy availability, distribution, and use in the twenty-first century, as has been defined through the marketplace and government policy, can only be solved in a hazy, complex, dynamic, and intertwined middle ground between: (1) science and philosophy; (2) private sector and government; (3) national, state, and local governmental entities; and (4) private citizens and corporate bodies (both governmental and private sectors).

CARBON-BASED FUELS: CURRENT AND FUTURE AVAILABILITY

Carbon-based fuels are often referred to as fossil fuels. The three commonly known carbon-based fuels are coal, petroleum, and natural gas. Under intense geologic pressure and the Earth's core heat, decayed and decaying materials—such as diatoms—primarily from the Carboniferous period (360–286 million years ago) formed organic compounds composed of hydrogen and carbon atoms. These hydrogen and carbon atoms are linked by chemical bonds to form hydrocarbon chains in one of two types of chemical formations (www.energyquest.ca.gov/story/chapter08.html, accessed April 20, 2004). Aliphatic formations are either straight or branched chains of hydrogen and carbon atoms, while aromatic formations involve something known as a benzene ring formation (see www.obio.com/hydrocarbon %20chains.htm, accessed April 20, 2004).

Natural gas is primarily methane. Gasoline fuel for use in cars contains a blend of liquid hydrocarbons. Coal is a complex aromatic hydrocarbon. In order to be made into gasoline or other useful products, fossil fuels such as crude petroleum must be separated into fractions through "cracking" (i.e., distillation). Petroleum refineries are large chemical processing plants where crude oil is "cracked," which reduces the molecular weight of crude oil hydrocarbons (www.towson.edu/~sweeting/enrich/petrolum.pdf, accessed April 20, 2004).

Following production of lighter fractions by cracking, various fuels and other petroleum products are separated. Gasoline is the most important and prevalent fuel produced from petroleum. Gasoline contains a mixture of liquid hydrocarbons. The amount of gasoline that can be derived from a barrel of oil is dependent on the quality of the crude petroleum, but roughly twenty-five gallons of gasoline can be refined from a barrel (forty-two gallons) of crude petroleum. In the internal combustion engine, oxygen molecules are reacted with gasoline to convert potential energy into kinetic energy. Combustion of gasoline produces exhaust gases, primarily dioxides.

COMMON FOSSIL ENERGY SOURCES:
U.S. SUPPLY AND USE

U.S. consumption of petroleum has risen steadily since 1950. Currently, the United States consumes approximately 20 million barrels of crude petroleum per day. The U.S. share of world crude petroleum consumption has fallen since 1950, but remained fairly stable since the early 1980s. As of 2005, the United States consumed approximately 25 percent of the crude petroleum produced on an annual basis. Perhaps a more interesting statistic, though, is the annual net imports as a percentage of U.S. crude petroleum consumption, which fell during the late-1970s due to the Organization of Petroleum Exporting Countries (OPEC) oil embargo, but has risen sharply since the mid-1980s. As of 2005, over 50 percent of the crude petroleum consumed in the United States was imported.

Crude petroleum supplies have declined significantly in the last century, as can be seen in figure 1.1. Due to accelerated petroleum production along Alaska's North Slope, total petroleum production in the United States rose in the mid-1980s to a level not seen since the early 1970s; since that peak period, however, petroleum production in the United States has declined by more than 30 percent in the last twenty years to approximately 1.9 billion barrels of oil per year. At the current rate of consumption, crude petroleum imports will have to rise significantly to meet unmet demand; thus, increasing further our dependence on foreign suppliers of crude petroleum.

U.S. consumption of petroleum and other fossil fuels has increased significantly in the last twenty years and shows no sign of declining. Crude petroleum imports have increased dramatically due to high demand and uncertainty about domestic supplies. Fossil energy, however, has by no means disappeared from the United States, and there remains an abundance of some hydrocarbons.

The most common fossil fuel in the United States is coal, which is found throughout the continental United States and is used as a heating source and for generating electricity. Coal can used in a solid form or made into coal gas—a highly volatile mixture of hydrogen and carbon monoxide. While crude petroleum consumption has increased dramatically in the last two decades, domestic coal production has increased at an even faster pace. Between 1980 and 2001, coal production in the United States increased by approximately 40 percent and the rate of increase remains steady.

Natural gas consumption has increased tremendously over the last two decades. Hailed as a cleaner alternative to heating oil and coal, many homeowners and businesses converted their furnace systems to natural gas. Natural

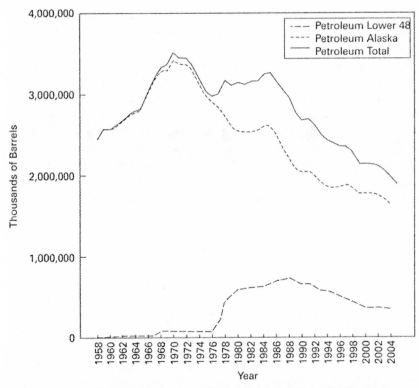

Figure 1.1.   Domestic Petroleum Extraction in the United States (1958–2005)

gas is primarily composed of methane gas but may also contain small quantities of other hydrocarbon gases, such as ethane and propane. Currently, the United States extracts and markets approximately 20 trillion cubic feet of natural gas on an annual basis.

According to the *U.S. Crude Oil, Natural Gas, and Natural Gas Liquids Reserves 2004 Annual Report*, there are 21.3 billion barrels of crude petroleum in proven reserves in the United States and U.S. protectorates, which includes approximately 165 million barrels of crude petroleum reserves discovered in 2004. Current production rates are approximately 8 percent of proven reserves per year. According to the U.S. Department of Energy (DOE), Energy Information Administration, as of 2004, the United States has 192.5 trillion cubic feet of natural gas in proven reserves, which includes a 2.6-billion-cubic-foot increase via new field discoveries and statistical adjustments. Production from natural gas reserves are growing at a rate four times the rate of new discoveries and represent approximately 4 percent of proven reserves per year.

The future is not bright for the supply of domestically available crude petroleum and other hydrocarbon energy in terms of the ability to meet demand. Even if new reserves exceed expectations and next-generation enhanced oil and gas recovery technology significantly improves recovery in mature oil wells and coal mines, shortfalls between supply and demand are extremely likely. The rate of consumption exceeds the combined totals of proven reserves and likely (based on immediate past discoveries) exceeds additions based on new discoveries. While the future is not known with absolutely certainty, science cannot tell policymakers and marketplace consumers that crude petroleum will be readily available much beyond the first century of the new millennium. Alternative or renewable energy source development is one method of increasing domestic energy supply for the twenty-first century and beyond.

CULTURE SHIFTS AND THE RISE OF GREEN POLITICS

The issue of fossil fuel availability provides impetus for asking the questions: Why pursue alternative energy and fuels, and why now? As a compelling "wicked" problem, energy policy in the twenty-first century requires the inclusion of a value-based discussion regarding the type of fuels and energy sources that are consistent with the political, economic, and social values important to consumers. Values are not uniform, nor are they spontaneously generated. Rather, values evolve over time and are a function of historically grounded, often region-specific trends (political culture) and present considerations (public opinion).

Values and beliefs shape the way society and individuals respond to energy policy issues. Costs, benefits, availability, and desirability are often evaluated in terms of preexisting values and beliefs. Values and belief systems tend to be fairly stable and frequently change at a glacial pace. At times, momentous events or circumstances are such that values and beliefs change far more quickly. The Great Depression, for instance, was a large-scale crisis that inspired cross-generational public support for the development of a social and economic safety net by government. World War II inspired a cross-generational commitment to the issue of national security and protection of basic human freedoms.

Shifts in long-established and widely shared values and belief systems (i.e., political culture) may change substantially along intergenerational lines. The most recent example of intergenerational change occurred in the 1960s and 1970s. In his book *Culture Shift in Advanced Industrial Society* (1990), Ronald Inglehart points to the Baby Boomer generation as substantially different from their parents' generation. Inglehart's analysis of longitudinal public opinion poll datasets supported his contention that earlier generations of Americans tended toward materialism. Materialists often to view public policy in terms of policy impacts on their own economic self-interest. Materialists frequently ask government to lower their tax burden and to limit the impact of social and economic maladies. Additionally, materialists call for government to fight inflation, which will negatively impact their economic well-being. Based on Inglehart's theory, materialists' support for energy policy would focus primary attention on readily available energy supplies and low prices.

The Vietnam War notwithstanding, baby boomers grew up in a period of relative luxury and surplus. Boomers were too young to be aware of the short-term post–World War II economic drop and employment concerns. Due in part to the relatively pacific socialization experiences of their youth, Ronald Inglehart argued that boomers have higher level needs compared to their parents' generation. In other words, basic social and economic needs were readily satisfied during the 1950s and early 1960s when boomers were in their youth. Absent concerns about basic needs, Inglehart argues that these postmaterialists are more concerned with higher level life issues, such as environmental justice and political and social equity, than they are with personal economic gain and tax relief.

In terms of alterative energy policy, postmaterialists see alternative energy as a method of improving the quality of life for human and plant and animal species on a global basis. Alternative energy production often costs more per kilowatt hour of electricity produced, but postmaterialists are more likely to view the increased cost as a trade-off necessary to promote and maintain global economic justice as well as freeing the United States from the yoke of international petroleum producers.

More particularly, boomer postmaterialists grew up at a time when environmental policymakers were considering policies to clean up the environment. As Inglehart (1990: 267–70) points out, postmaterialists worked to advance the issues of social and environmental justice in the 1960s and 1970s. Despite Reagan's efforts in the 1980s to revitalize traditional crude petroleum and nuclear energy policies, interest in alternative and renewable energy sources remained. Cultural values regarding the environment and the need to pursue energy independence through new methods of power generation were not abandoned, although national laboratories working to develop alternative energy systems were significantly underfunded.

THE RISE OF GREEN POLITICS AND ENVIRONMENTAL PUBLIC INTEREST GROUPS

The political culture shift of the 1960s and 1970s led to the rise of the greens. In European politics, the Green Party became a major force in the politics of many Western European governments. In the United States, it is more applicable to refer to the green movement, since the party's direct influence is very limited. The green agenda is based in environmental activism seeking to create a balanced relationship between human society and the protection of flora and fauna. Greens tended to reject industrial capitalist society, arguing that industrialization created an unbalanced and unjust relationship between human society and the natural environment. At the same time, greens tended to reject the other dominant world view of the time—namely, Marxian ideology—in large part because Marxism focused primarily on the needs of human society, which, in practice, tended to produce similar violations of environmental justice as occurs in industrial capitalist societies.

The greens have, as part of their political agenda, the development of alternative energy policy. Petroleum and other hydrocarbon fuels are rejected by greens because of damage to the environment caused by emissions. From their postmaterialistic perspective, greens view zero- or low-emission alternative and renewable energy sources as representative of the true costs of energy, whereas petroleum and other hydrocarbon fuels, while initially less expensive, actually have significant hidden costs in the form of emissions that harm the environment as well as creating inequalities in human society through emissions-related disease and mortality.

A major impact of the political culture shift and the rise of green ideology that occurred in the 1960s and 1970s can be seen in the first major statutorily derived steps toward a new energy paradigm. Emergent environmental and energy policies were given life by youthful postmaterialists who staffed agencies at the federal and state levels, working for the advancement of environmental quality through efforts to protect land, water, and air quality in a holistic manner. The dream of reduced pollutants from hydrocarbon fuels remains alive in the work and values of postmaterialists (Holland et al. 1996; Toke 2000).

A secondary, but equally important, impact of the shift toward postmaterialist values can be seen in a new political and social activism that has become an increasingly acceptable method of advancing political and social causes. The 1960s and 1970s witnessed the growth of public interest groups—pressure groups dedicated to generalized societal benefits rather than to narrowly defined economic benefits for group members. Postmaterialist values and public interest groups often have parallel value structures rooted in the desire to serve the collective good. In tandem, postmaterialists' commitment to using activism as a method of promoting social and political causes, and public interest groups' sense of social responsibility served to blunt the impact of the Reagan administration's efforts to promote energy "independence" through aggressive domestic petroleum exploration on U.S. public lands as well as efforts to advance nuclear power as a solution to energy dependence and future short-falls in petroleum and other hydrocarbon fuel sources.

The aforementioned question of "Why alternative energy?" is answerable on both an empirical and normative basis. Empirically, we know that fossil fuels are nonrenewable sources of energy and that they are being depleted at a fairly rapid rate. Normatively, it is apparent that the postmaterialist view of social and economic justice leads many individuals to consider alternative energy to be well worth the cost of promoting social and economic justice and energy independence. Charting historical trends in fossil energy consumption and supply provides the empirical basis for answering the question: Why now?

NEW ENVIRONMENTAL PARADIGM AND ALTERNATIVE ENERGY

Emerging over a decade prior to the work in Inglehart (1990), the New Environmental Paradigm (NEP) was one of the first theoretical developments to recognize a value shift that impacted the way in which individuals viewed human society in relation to the environment. The NEP does not speak directly to the issue of alternative energy, but it does serve as a basis for understanding the historical forces that have led to the development of alternative energy policies. In essence, the NEP argues that movement toward a new view of environmental justice and growing interest in less environmentally damaging human interactions with the natural world are not based on marketplace rationality and the desire for cheaper energy sources. Rather, the direct cost of electricity generated from green energy invariably costs more per unit than does electrical energy generated from fossil fuel.

The New Environmental Paradigm

1. Human beings are but one species among the many that are interdependently involved in the biotic communities that shape our social life.
2. Intricate linkages of cause and effect and feedback in the web of nature produce many unintended consequences from purposive human action.
3. The world is finite, so there are potent physical and biological limits constraining economic growth, social progress, and other societal phenomena.

Source: Catton and Dunlap 1978: 45.

The demand for alternative energy is not based on desire for a cheaper source of energy, since alternative energy remains more expensive than energy produced from fossil fuels; rather, the demand for green power or nonhydrocarbon alternative energy sources is part of a cultural or sociological shift.

INSTITUTIONAL CHANGE AND INFLUENCE

Traditionally, political scientists and policy analysts have looked to institutionalism—the study of the rules, processes, and structure of political institutions—to describe, explain, and predict policy choices and outcomes. A more advanced way of thinking about institutions and institutional change is embodied in the general theory known as new (or neo) institutionalism. New institutionalism is a sophisticated methodology that allows political scientists and policy analysts to study both institutional characteristics and constraints while simultaneously studying the values, opinions, and beliefs of institutional actors making decisions within particular political institutions. In this case, the institutions being addressed include: Congress, the presidency, the courts, and the public bureaucracy. Changes that have been wrought in our political institutions help answer the central question of this chapter, namely "Why alternative energy and fuels?" New institutionalism helps conceptualize the combined impact of empirically-generated data regarding energy demand and supply as well as the impact of value shifts that have been incorporated into our collective vision for marketable and policy-based energy solutions that reach standards of political and social acceptability.

As an institution, Congress changes very slowly. The basic structure and rules governing legislative process are not dissimilar today from the structure and rules of eighteenth-, nineteenth-, or twentieth-century Congresses. Changes that have occurred in Congress tend to be related to party control of the institution and the changing political and social values, beliefs, and opinions of congressional leaders and membership. Over the last three decades, there have been at least two major shifts in congressional leaders and membership that may help us answer the chapter's central questions. The two major shifts occurred in: (1) the 1974 general election, and (2) the 1994 midterm congressional election.

The 1974 election is well known for the Watergate scandal in which individuals closely tied to President Richard Nixon broke into the campaign offices of Nixon's Democratic rival, George McGovern. While a major event in U.S. political history, the scandal overshadows other major developments that emerged from the election. For many baby boomers, the 1974 election was their first major involvement in shaping the U.S. political landscape and for many of these young postmaterialists, the election turned on issues important to their generational cohort. The Democratic Party retained control of Congress, and the freshman class of congresspersons was very large and powerful. The Democratic Party gained nearly fifty seats in the House of Representatives and four seats in the Senate. The new 94th Congress was composed of many young members, who broke from institutional traditions of apprenticeship and deference to congressional leaders and norms to pursue policy items that were important to them and to their postmaterialist constituencies.

Several landmark acts of environmental legislation had already been passed earlier in the decade, and the 94th Congress helped to preserve and expand the spirit of those laws. Within a half decade, congressional legislation led to the creation of National Renewable Energy Laboratories (NREL). With the development of NREL, existing public and private laboratories working in the area of energy policy witnessed increased funding and an expanded charge to advance U.S. energy policy and explore new alternatives. Congressional legislation and oversight functioned to steer policy toward cleaner energy and improving technologies associated with alternative and renewable energy sources.

Although the Republican Party gained control of the U.S. Senate in 1981 and retained control until 1987, the Democratic Party kept its thirty-plus-year control of the House of Representatives. Divided party control of Congress served to protect alternative energy policy from termination, but funds for these policies were severely constrained. Now senior members of the House of Representatives, the Democrats of the freshman class of 1975 remained committed to alternative energy policy. The basic values and beliefs of these congresspersons were still present both personally as well as in the spirit of nearly two decades of legislation that had emerged during their tenure in office.

The congressional midterm election in 1994, however, led to another major change in congressional leadership and membership as Republicans swept into office with majorities in both the House of Representatives and the Senate. Budgeting for the research and development of alterative energy sources[1] such as wind, solar, and geothermal power decreased, followed by a smaller decrease during the 105th Congress. However, with President Clinton's reelection in 1996, federal funding for alternative energy research and development increased until the passage of President George W. Bush's budget in 2002. More specifically, the largest decrease was in funding for solar energy research, while funding for wind power fared somewhat better. Nevertheless, new institutionalism, which combines an understanding of institutional processes and rules governing decision making with an understanding of the individual members' values and beliefs, may offer some insight into the question of "Why alternative energy?" as well as the evolving commitment to energy policy issues by Congress.

The evolving nature of the presidency and the priorities of individual presidents offer important insight into the issue of alternative energy. Presidents are often policy leaders, establishing national priorities and broad goals for their terms of office and for the future. President Theodore Roosevelt, for instance, demonstrated prescience when he advocated the establishment of the national park system, reserving public lands for the recreation of citizens as well as preserving native plant and animal species in a largely pristine environment. Other presidents have demonstrated similar concern with the preservation of the environment and the need to balance societal needs with a responsibility to be good stewards to the natural environment.

President Nixon signed major legislation—the National Environmental Policy Act of 1969 (NEPA), 42 U.S.C. 4321–4347; the Clean Air Act of 1970 (CAA), 42 U.S.C. ss/1251 et seq.; the Endangered Species Act of 1973, 7 U.S.C. 136 & 16 U.S.C. 460; and the Occupational Safety and Health Act of 1970, 29 U.S.C. 651—establishing a national commitment to environmental protection. The legislation moved national energy policy toward a greater commitment to alternative energy; primarily for the purpose of reducing harmful emissions resulting from the combustion of hydrocarbon fuels. President Jimmy Carter signed into law the Clean Water Act of 1977 (CWA), 33 U.S.C. ss/1251 et seq. The landmark laws—detailed later in this book—reflect a growing postmaterialist value shift during this period, recognizing that the economic and social needs and desires of humans must be carefully balanced with the needs of the natural environment.

Of more specific importance to the issue of alternative energy, President Carter established the DOE as a cabinet-level office in the Executive Office of the President. Previously, energy issues were addressed by various federal offices. The creation of DOE was the recognition of the need to better coordinate national energy needs with energy resource reserves and production. The DOE conducts an annual energy policy review, determining the current status of fuel availability and use as well as coordinating external and internal research funding for the development of energy sources consistent with the goals of environmental policy commitments: that is, reduction in fuel emissions and fuel efficiency.

The Carter presidency witnessed two major crises in the nation's energy paradigm. First, Carter dealt with an oil shortage resulting from a decision by the Organization of Petroleum Exporting Countries (OPEC) to cut oil production, thus driving up prices and reducing supply. The oil embargo demonstrated the dependency of the United States and other industrialized nations on foreign oil supplies. If supplier nations were displeased with the United States, then organizations such as OPEC could use the oil supply as a diplomatic weapon to influence U.S. policy. Second, nuclear power as an alternative fuel source was dealt a major blow when a serious nuclear power plant accident occurred at Three Mile Island in 1979, releasing radiation into the atmosphere. Fortunately, the nuclear accident was contained, but the incident demonstrated the potential dangers of nuclear power plants.

President Reagan did not promote renewable energy or alternative energy sources. Rather, he unsuccessfully attempted to advance nuclear energy. Reduced funding for alternative energy slowed the development of viable alternative energy systems. The 1980s was the beginning of an accelerated commitment to petroleum, natural gas and coal as primary sources of energy.

A major shift toward alternative energy development occurred during the last year of the George H. W. Bush presidency. In 1992, Congress passed the Environmental Policy Act (EPAct), which authorized the Environmental Protection Agency (EPA) to establish more stringent air quality standards. While President Bush did not have sufficient time in office to fully implement EPAct, the law served as an important policy shift that was favorable to the development of low- or zero-emission alternative or renewable energies. EPAct provided an early momentum for the Clinton administration's energy policy agenda.

As a policy leader, President Clinton used his presidency to establish a renewed commitment to environmental and alternative energy policy. As mentioned earlier, the Republican takeover of Congress in the 1994 midterm election produced divided government and some friction over energy policy. Funding for alternative energy followed a growing budgetary commitment during the presidency of George H. W. Bush. In other words, there was a policy momentum that had already been established before Clinton emerged as a national policy agenda setter. During Clinton's presidency, funding for alternative energy research and development—for the first time in nearly thirty years—reached parity on an adjusted dollar basis with funding levels of the mid-1970s.[2]

As a policy leader, Clinton, the first baby boomer elected president, astutely recognized rising postmaterialist values, the growing prominence of green politics, and the sociological trends associated with the NEP. The Clean Cities Program, a programmatic response emerging from statutory mandates found in EPAct of 1992, was implemented by the Clinton administration and authorized by Congress in 1993. The program was designed to develop partnerships between national, state, and local governments, encouraging and rewarding state and local efforts to reduce harmful emissions and to improve air quality. As a former governor, Clinton recognized that innovative forces at the state and local levels will most likely produce unique policy strategies needed to move from the fossil fuel economy toward a twenty-first-century energy paradigm based on low- or zero-emission alternative and renewable energy sources.

The current presidency of George W. Bush began by echoing a Reagan administration proposal to open the Arctic National Wildlife Refuge (ANWR) to petroleum and natural gas exploration. As a candidate for the presidency, George W. Bush saw ANWR as one solution to U.S. dependence on foreign petroleum supplies. The proposal was challenged by Democrats in Congress and by environmental public interest groups. At this point, ANWR remains closed to petroleum or natural gas exploration. Following the events of September 11, 2001, George W. Bush used his role as a policy leader to establish a renewed commitment to alternative energy policy development. The president proposed and Congress passed the Hydrogen Initiative. The president views hydrogen as a realistic fuel of the future. With the notable exception of ANWR petroleum and gas recovery, many of Bush's energy policy initiatives are incorporated in EPAct of 2005.

The federal courts have played a role in the momentum shift toward alternative energy development. The changing role of the courts in legitimizing an expanded role of the national government in policy areas such as energy policy can be traced back to institutional changes in the 1930s. The courts have reinforced the legitimacy of statutes and administrative rules governing the application of environmental policy related to NEPA, EPA, CWA, and CAA. By establishing the legitimacy and enforcement of such regulations, the courts have allowed for stringent applications of the law with the ultimate intent of promoting better stewardship of the environment and balancing societal needs with the needs of nature.

The Supreme Court has legitimized the national government role in creating and enforcing environmental quality policies. The recent U.S. Supreme Court opinion in *Engine Manufacturers Association, et al. v. South Coast Air Quality Management District, et al.* (2004) illustrates the Court's current stance on the role of the national government in shaping environmental and energy policy, in this case through the enforcement of uniform air quality standards. In this instance, the Court viewed Los Angeles's requirement for bus and truck operators to use only locally approved low emission vehicles for transportation of individuals and goods within the county area as a preemption of the national government's regulation of interstate commerce. The Court has used judicial review to protect the national governments' power to enforce nationwide standards for environmental quality, which ultimately leaves it to the national government to create incentive structures to develop alternative energy through regulations limiting the use of hydrocarbon fuels or at least through tightening emissions standards. While some environmental public interest groups might view the Court's majority opinion as a "loss" because the justices largely rejected an aggressive local government proenvironment policy initiative, in the long term the legal opinion protects and likely bolsters the national environmental policy agenda.

The public bureaucracy plays a very significant role in shaping alternative fuel development and use. Institutional changes in the public bureaucracies were a function of those bureaucracies coming into existence as a result of aforementioned legislation of the late 1960s and early 1970s. The DOE and EPA (an independent regulatory agency) were created during a period of growing environmental awareness and rising postmaterialist values. The new agencies, whose employees are frequently advocates for environmental protection, use the power of regulation to restrict the use of hydrocarbon fuels *if* the combustion of the fuel produces unacceptable levels of emissions into the environment. In recent years, particularly during the Clinton administration, emission standards were tightened through the EPA's National Ambient Air Quality Standards (NAAQS) (which emerged from the CAA of 1990, renewed). The NAAQS requirements will regulate ozone concentrations in the United States, which will have the effect of shaping automobile manufacturing decisions and the development of cleaner burning internal combustion engines and, ultimately, lead to a greater use of alternative energy sources that produce lower levels of emissions. The emission standards of the NAAQS will go into effect in 2010. The question of "Why alternative fuels?" may be answered quite simply because at least some alternative energy sources (e.g., hydrogen fuel cell vehicles) are less likely to contribute to air quality problems, reducing the likelihood of violations of federal air quality standards.

Regulation, however, is only one way in which public bureaucracies shape the development of alternative energy in the United States. Since the 1970s, the DOE has invested billions of dollars in alternative and renewable energy systems through the Office of Energy Efficiency and Renewable Energy (EERE) (www.eere.energy.gov, accessed April 20, 2004). Through grants to university researchers and contracts with private sector enterprises, the EERE has brought government investment to bear on the development of alternative and renewable energy systems. Grants to universities have involved the development of demonstration projects, which illustrate the viability of alternative energy applications as well as identifying areas where improvements could be made. Grants have also contributed to scientific discoveries that make alternative energy less expensive and more viable for the average consumer. Additionally, DOE's national laboratories, such as Argonne National Laboratory and the National Renewable Energy Laboratory (NREL) work with alternative energy industries to produce technologically advanced alternative energy systems. While Argonne National Laboratory has been in existence since the 1940s, NREL emerged from the newly created DOE in the mid-1970s. Through NREL, the DOE has developed demonstration projects in many developing nations with the intent of illustrating the utility of alternative energy in suboptimal settings.

U.S. Department of Energy National Laboratories

Argonne National Laboratory
Brookhaven National Laboratory
Idaho National Engineering and Environmental Laboratory
Lawrence Berkeley National Laboratory
Lawrence Livermore National Laboratory
Los Alamos National Laboratory
National Energy Technology Laboratory
National Renewable Energy Laboratory
Oak Ridge National Laboratory
Pacific Northwest National Laboratory
Sandia National Laboratory

Source: U.S. Department of Energy, 2005.

The DOE has also developed partnerships with automobile manufacturers to promote energy independence. The Freedom CAR and 21st Century Truck programs are intended to bring manufacturers and government together to develop new alternative energy vehicles that are powered by hydrogen fuel cell technology.

The public bureaucracy provides solid evidence for new institutionalism. The effect of a value shift in U.S. postindustrial society combined with major institutional change shaped the regulatory and distributive policy choices of public bureaucracies. Of particular importance, the policy choices favored and continue to favor the development of alternative and renewable energy.

GLOBAL DEMANDS AND CONFLICT

Global demand for limited petroleum resources is another important explanation for the development of alternative or renewable energy resources. In the last thirty years, the world has changed dramatically. Nations, such as China, which were once considered underdeveloped nations with stagnant or authoritarian economies, are industrializing at a rapid pace. China's consumption of crude petroleum increased by 29.1 percent between 1999 and 2003, which accounted for 37 percent of the total increase in world petroleum demand during that same time period. Developing countries account for approximately 78 percent of the increased demand for petroleum between 1999 and 2003.

Increased global demand may not directly impede the ability of the United States to meet its petroleum needs, but a reduced world supply of crude oil will likely mean increased world prices for petroleum.

Supply might also become more limited due to international conflict, which could disrupt petroleum distribution or lead to imposed oil shortages by OPEC and non-OPEC nations supplying oil to the world market. During Operation Desert Storm, retreating Iraqi armies were ordered by Iraqi president Saddam Hussein to ignite oil wells as the armies ended their occupation of Kuwait. In the recent Iraq War, partisans in occupied Iraq have ignited their own nation's oil wells in protest of the U.S. occupation of the nation. In both instances, oil production slowed and world oil prices rose. Additionally, the conflict in Iraq has led OPEC nations to slow oil production to make their displeasure known with regard to U.S. military involvement in the Middle East.

Changes in global demand and international conflict invite policymakers to more actively consider the development of domestic alternative or renewable energy sources. Given the level of demand on energy in the United States (as of 2003, the United States consumed approximately 25 percent of all petroleum produced worldwide), it is unlikely that alternative or renewable energy sources will eliminate the need for oil imports, but the development of these additional energy sources would likely reduce U.S. economic and political vulnerability to changing market forces as well as international efforts to use petroleum as a diplomatic wedge issue against U.S. foreign policy efforts.

CHAPTER SUMMARY

There is no simple answer to the question, "Why alternative energy?" Multiple forces influence responses to the question. To a significant degree, responses are driven by empirical evidence—what is known (or thought to be known) about the availability of petroleum, which is the foundation of the world economy and the centerpiece of the dominant energy paradigm. Cultural and political changes that have occurred over the last thirty to forty years play a tremendous role in advancing the alternative energy paradigm and have made resources available to develop technically and economically feasible solutions to the eventual decline in petroleum-based economics.

Having addressed this issue of "Why alternative energy?" it is only natural to then consider the issue of "How?" Alternative energy is here to stay and it will play an ever-increasing role in meeting our energy needs. Primary reliance on petroleum as an energy source may not be feasible within two or three generations. Answering the question of "How?" becomes particularly important if our demands on energy are going to be met in the years ahead.

Tremendous time and effort will be required to make the alternative energy paradigm a politically and socially legitimate and economically feasible solution to the nation's energy needs.

NOTES

1. I used NREL's budget as a proxy for budget commitment to alternative energy research and development funding.
2. As with earlier discussion of funding for research and development for alternative energy, I am using NREL budget data.

WORKS CITED

Anderson, James. 1984. *Public Policymaking: An Introduction*. New York: Houghton Mifflin.

Bellah, Robert, Madsen, Richard, Tipton, Steve, Swidler, Ann, and Sullivan, William. 1992. *The Good Society*. New York: Knopf.

Berinstein, Paul. 2001. *Alternative Energy: Facts, Statistics, and Issues*. Westport, CT: Oryx Press.

Berk, Richard A. 1981. *Water Shortage: Lessons in Conservation from the Great California Drought, 1976–1977*. Cambridge, MA: Abt Books.

Blackwood, John. 2002. *Energy Research and the Cutting Edge*. New York: Nova Science Publishers.

Catton, Bruce and Dunlap, Riley. 1978. Environmental Sociology: A New Paradigm. *The American Sociologist* 13 (February): 41–49.

Douglas, Mary and Wildavsky, Aaron. 1973. *Risk and Culture*. Berkeley: University of California Press.

Etzioni, Amitai. 1988. *The Moral Dimension: Toward a New Economics*. New York: Collier Macmillan.

Hardin, Garrett. 1968. The Tragedy of the Commons. *Science* 162(1968): 1243–48.

Heilbroner, Robert. 1994. *21st Century Capitalism*. New York: Norton.

Holland, Kenneth, Morton, F., and Galligan, Brian. 1996. *Federalism and the Environment: Environmental Policymaking in Australia, Canada, and the United States*. Westport, CT: Greenwood.

Inglehart, Ronald. 1990. *Culture Shift in Advanced Industrial Nations*. Princeton, NJ: Princeton University Press.

Mills, C. Wright. 1967. *The Sociological Imagination*. Cambridge: Oxford University Press.

Simon, Christopher A. 2007. *Public Policy: Preferences and Outcomes*. New York: Longman.

Toke, David 2000. *Green Politics and Neo-Liberalism*. New York: St. Martin's Press.

Van Beuren, Ellen, Klijn, Erik-Hans, and Koppenjan, Joop. 2003. Dealing with Wicked Problems in Networks: Analyzing and Environmental Debate from a Network Perspective. *Journal of Public Administration Research and Theory* 13(2): 193–212.
World Bank 2006. "World Development Indicators Database." siteresources.worldbank .org/DATASTATISTICS/Resources/GDP_PPP.pdf, accessed October 19, 2006.

COURT CASE

Engine Manufacturers Association, et al. v. South Coast Air Quality Management District, et al. (2004).

WEB SITES

www.energyquest.ca.gov/story/chapter08.html, accessed April 20, 2004.
www.obio.com/hydrocarbon%20chains.htm, accessed April 20, 2004.
www.towson.edu/~sweeting/enrich/petrolum.pdf, accessed April 20, 2004.

Chapter Two

Studying Public Policy and Alternative Energy/Fuels

INTRODUCTION

Social values and economics shape the debate over optimum energy solutions. In public policy, consumers have an active voice in the production process, requiring an understanding of many technical issues. Technical factors related to public policy are often unknown or not understood by the average citizen stakeholder. Absent awareness or understanding, the average citizen is limited in his or her ability to participate in policy dialogs and decision making.

Complicating matters further, technical information is frequently not agreed on by policy experts. The meaning and value of data and solutions based on these data are often closely tied to assumptions and values underlying collection and interpretation. For instance, oil drilling firms make certain assumptions about consumption patterns and petroleum deposits when predictions are developed related to the relative abundance of product, as well as the costs of extracting petroleum (costs that may or may not accurately reflect the total costs of production—for example, potential harm to the environment—and opportunity costs to society when extraction occurs on public lands). The same problems arise when alternative energy engineers and researchers coalesce to discuss new generation energy systems. In either case, the uninitiated would likely find the asymmetric communication confusing and uninformative in terms of the decision-making process.

Technical issues are often closely tied to competing values, particularly in the case of energy policy. Energy solutions might at first glance seem quite reasonable. Once further information is provided and considered in relation to basic values, solutions might be unacceptable. Sour heavy crude, for instance, is abundant and relatively cheap, but unrestrained use of this petroleum product yields highly toxic sulfur emissions that would reduce air quality and contribute to acid rain.

Advocates of alternative energy solutions often point to wind turbines as a method of meeting energy demands of the future lacking the nasty impacts of petroleum solutions such as greater use of heavy sour crude petroleum. But wind turbines have other impacts that might be distasteful for some individuals. For example, individuals might find the turbines aesthetically unpleasant. Although improvements have been made in new-generation turbine design, avian mortality can occur as well as sound disturbances as a result of turbine operation.

In the examples above, the choice could be described as sulfur emissions versus aesthetic impacts. The point here is that with any policy decision, particularly one in which scientists disagree about technical issues and in which values play a large role, there are a multitude of choices that must be weighed carefully. Each choice involves trade-offs or the associated costs and benefits of making one choice as opposed to another. In this chapter, I will outline several aspects of the policy process that shape the way in which choices are made.

ROLES FOR PUBLIC POLICY IN
ALTERNATIVE ENERGY/FUEL DEVELOPMENT

Public policy is often legitimized along two dimensions: (1) economic considerations associated with a good or service, and (2) issues related to individual property rights. In economic terms, a good or service can be analyzed along two dimensions: considerations of rivalrousness and considerations of excludability. Rivalrousness relates to basic economic theories regarding price, supply, and demand. In essence, if a good or service is relatively limited in supply and demand for the good is appreciable, the price of the good-will increase to a point of equilibrium, a point at which supply meets demand. Rivalrousness means that ability to pay for a good or service determines who gains access; short supply and high demand means that prices will likely increase, leaving potential consumers in the position of having to pay a particular price in order to gain access or ownership to a good or service.

The spatial diagram (figure 2.1) illustrates the issue of rivalrousness in terms of gasoline supply and demand. In this instance, gasoline demand is fairly inelastic—demand does not respond too much to issues of price, which

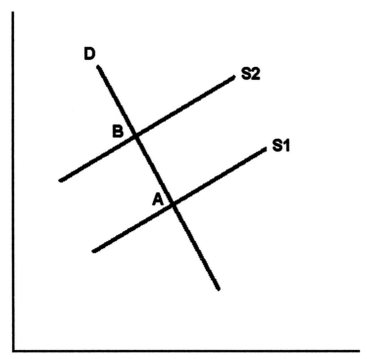

Figure 2.1.   Rivalry and the Market for Gasoline

is a function of supply and demand. Rich or poor, people who drive need gasoline to get to work and to go shopping. If supply moves from S1 to a lower level of supply at S2, then price for gasoline increases. As you can see, the poor person may not be able to afford the gasoline. Excludability means that whomever can afford the product will gain access to it; if it is beyond an individual's budget constraint, then it is not obtainable no matter how much it might be needed. The poor person, in this case, would have to figure out a method of moving their budget constraints beyond point B (assuming he or she wants to eat as well as purchase gasoline).

Excludability refers to the idea of ownership of some type. The owner of a house has excludable rights to her or his home. While she or he owns the house, nobody else can own the house or use the home legally without her or his express permission or without a legal contract of some type. Excludability becomes a very important subject in the United States, largely because the social contract is grounded in the assumption of individual property rights. Liberal democratic traditions generally assume that individuals have pledged themselves to the protection of one another's rights; property and/or otherwise.

There are certain goods that are neither excludable nor rivalrous. In other words, multiple parties can access the goods or services simultaneously (i.e., nonexcludable) and the forces of marketplace capitalism are less likely to significantly influence the price of the good, largely because the good is not easily divided into tangible dimensions (i.e., nonrivalrous). Goods of this nature are often referred to as public goods. A good example of a public good is air. Multiple individuals breathe at the same time and air is not easily divided into marketable quantities, and, given the forces of supply and demand, one could easily deduce that price would approach zero. Frequently, government manages public goods to ensure equitable distribution and protection of the good for current and future use (see Ostrom 1990).

Other goods (e.g., water, energy, public health) are not so easily defined in terms of their relative excludability or rivalrousness. There is a tremendous debate within public policy regarding the relative nature of goods and services offered through government. Neoclassical economists often argue that many of the goods or services provided by government are not pure public goods, but are, in fact, marketable private or semiprivate goods (see Hayek 1944). Conversely, other individuals find that the use of many goods or services impacts the property or basic human rights of other individuals, thus necessitating government intervention in the distribution and use of such goods or services. This more "progressive" approach to public policy ties a discussion of rights into a discussion of economic theory (see Michelbach et al. 2003; Rawls 1971).

The role of public policy is decided on these two major dimensions: the economic nature of the good or service and the right-based aspects of the good or service. Energy policy has, in many respects, returned to an earlier position as a focus of policy debate. Historically, energy policy was a function of individual or communal choices. Energy was often viewed as a communal good. Energy demand was fairly limited and individuals could easily gain access to it. For example, wood from a nearby forest or surface-exposed crude petroleum tars were used for cooking and for heating purposes in families or tribes (Bowers 2004). The same was true when human beings moved toward a more urbanized model of living, relying on a community grist mill to turn their grain crops into flour or meal for consumption or sale (see Unger 2004). Decisions about energy policy were made at a fairly low level and market-based economic forces of rivalry or excludability played almost no role in energy generation or use decisions.

As modern industrialized societies emerged, energy demands grew tremendously. While supply was fairly plentiful, access to supply was often a function of transportation of stored or produced energy to cities and the distribution of the good among consumers. As prices of energy production, storage, and transportation declined, energy came to be viewed as a marketable public good. Ceteris paribus, the energy market moved toward inelasticity for two reasons: (1) prices were relatively low and likely to remain so; and (2) the modern economic and social paradigm required energy consumption in order to exist. For much of the nineteenth and all of the twentieth centuries, this model drove energy policy—a private policy rather than a public policy. If anything, government assumed a primary role in meeting demand and reducing individual cost by increasing supply through distributive policies related to capital projects—for example, dam, power line and pipeline easements, and road construction.

At the dawn of the twenty-first century, energy supply issues abound and government's ability to enlarge supply through public policy faces new challenges. Additionally, there is a renewed debate over the appropriateness of energy policy as a private market-based policy area, which will probably become a louder debate if energy supplies become constrained significantly. As domestic supplies have declined and as supply shocks began to occur in the 1970s, government was called on to solve the crisis through increasing supply and reducing costs—solutions unworkable because the former cannot easily and quickly be achieved and the latter violates the basic laws of economics in a manner that does not efficiently manage supply constraints.

For individuals who reject the market mechanism, the notion that the marketplace could and would appropriately fuel the needs of a modern society was further scrutinized.[1] Given the heavy reliance on energy to exist in the modern society and to use individual property, many critics of private market-based solutions argued that basic rights and needs made energy a public good. From the public good perspective, government must manage energy resources and accessibility. These same critics argue that other rights and goods were being negatively impacted by the market forces and choices related to energy. Fossil fuels, for instance, were impacting public goods such as clean air, causing public health crises for individuals and groups of individuals living within communities. In order to preserve other public goods, it was argued, government must play a role in shaping the energy marketplace. Proposals to expand supply were largely built on assumption that increased supply would mean *lower energy costs* and *greater accessibility*. While supply might be increased, cost reductions and accessibility may not come to fruition due to other factors.

POLICY PROCESS

It is unlikely that there will be a purely private or a purely public solution to impending energy supply shortfalls. Private energy solutions are hampered by public cynicism about energy firms. Media reports and court scrutiny in high-profile cases involving Enron and Haliburton have potentially damaged citizen confidence in the energy industry. Private solutions emerge only as fast as resources are committed to research and development programs for new technologies. If a firm sees more revenue in existing technology and processes, it is unlikely that it will disturb the status quo. Conversely, firms will initiate a search process for new business strategies with the expectation of greater revenue and profit margins. Purely public solutions are slow in emerging and are often the result of statutory or common law requirements, public-private partnerships, or government research grants or demonstration project funds to promote private investment in next generation technology and infrastructure. There is a high probability that public policy will play a significant role in alternative energy solutions generated both publicly and privately. For this reason, it is important to understand the process by which policy is created.

Public policy is a phrase thrown about quite often. In everyday discussions at the office or at home, the phrases "There ought to be a law..." or "Government ought to do something..." are not uncommon.[2] In a public policy class, perhaps an environmental policy class, interest in completing these phrases with some form of empirically generated policy is often what attracts students to college courses. If, for instance, a student thinks that solar panels and wind vanes are the methods by which government ought to pursue a greater energy supply, then it is quite likely that the person will be an ardent advocate of alternative energy policy. Alternatively, a student might find exploration of nuclear energy to be the optimum policy solution and will advocate the building of reactors across the United States. Either way, of course, the policy positions will provide great classroom banter, but will not move public policy from X to Y, so to speak; in essence, simply advocating a policy (X) does not mean that policy will come to fruition (Y). There is a whole policy *process* that occurs that may explain *why* some proposals are seriously considered and move along to the point of becoming actual public policy while other—possibly good—ideas simply flounder. The process of policy also explains *why* and *how* policy ideas emerge as public policy and perhaps helps one understand potential policy *outcomes*.

In his book, *An Introduction to the Study of Public Policy* (1970), Charles O. Jones outlines the five major steps in the policy process: (1) agenda setting, (2) policy formation, (3) policy implementation, (4) policy evaluation, and (5) policy termination or change. In most cases, policy does not terminate, which means that the process is cyclical rather than a linear process. In other words, there are always future opportunities to advance ideas that have been rejected on their first go-around, and it is also quite feasible to alter existing policies to meet new policy preferences.

Agenda Setting

The first step in public policy is for an issue to become a part of the policy agenda. A multitude of interests, issues, and policy solutions are continually clamoring to become a part of the policy agenda, but amazingly very few actually weave their way through the process and become seriously considered topics or options actively discussed by political executives, legislators, judges, and public administrators. Becoming a part of the policy agenda is often a function of circumstances within political institutions and conditions in the domestic and international community. The president plays a tremendous role in shaping the policy agenda through his unique position as the only nationally elected leader.[3] In the 2000 general election, George W. Bush and Albert Gore Jr. presented two very different policy agendas for energy policy; the former was interested in opening the Arctic National Wildlife Refuge to petroleum exploration, while the latter candidate promoted alternative clean energy policies (see *Nation's Cities Weekly* 2000). Congress represents the values of the nation as well as the parochial values of districts and states. Often, there are a great variety of priorities brought forth by legislators representing their districts or states, which impact the policy agenda. Of course, political party identification of the president and majority of Congress, in either or both chambers, reflects value preferences and subsequently impacts the policy agenda (see Binder et al. 1999; Krehbiel 1993). Although not overtly political, the federal judiciary is appointed by the president with the advice and consent of the Senate. Presidents are often drawn to individuals who share their political and social values; thus, the courts are influenced by political preferences and will shape the policy agenda accordingly (see Hulbary and Walker 1980). For example, judges appointed by Democratic presidents in the 1950s and 1960s shaped, and continue to shape, the environmental policy agenda through court decisions.

Circumstances in the domestic (Kingdon 2003) and global (see Weatherford 1988) social and political environment also impact the policy agenda. One of the most obvious circumstances is a crisis. As mentioned in chapter 1, an oil crisis in the 1970s led to shortages of supply and skyrocketing prices at the gas pump. In the 1970s case, the Organization of Petroleum Exporting Countries cut production of oil to drive up prices and profits. While other nations' citizens were used to paying high prices for fuel, domestic reaction in the United States was one of shock and dismay. Limited energy supply meant higher prices and excess energy demand that could not be met under the circumstances. There were angry calls for something to be done to meet energy demand. Consequently, energy policy became a prominent part of the policy agenda, with alternative energy becoming a more prominent policy solution.

In the 1980s and 1990s, energy policy was a prominent item on the policy agenda. Efforts to promote green energy solutions remained in the background, often quietly explored by national energy laboratories. Falling oil prices—prices that hit record lows in the late 1990s—as well as discussion of natural gas as an almost inexhaustible domestic energy resource overshadowed alternative energy policy. While major discoveries of natural gas have been made recently (e.g., approximately 50 trillion cubic feet of natural gas under Fort Worth, Texas), it is now generally thought that cheap petroleum and natural gas domestic supply is on the decline. In 2003, Secretary of Energy Spencer Abraham indicated that natural gas supplies are declining more rapidly than expected, necessitating the large-scale importation of natural gas from other nations (see Cook 2003).

In the early twenty-first century, the nation has witnessed a return to higher petroleum prices; but this time, the cost factor is a function of things other than production decisions per se. The world economic outlook has changed in the last several decades. Nations such as India and China are rapidly moving toward more-developed-nation status. A middle class has formed in many nations, which means greater demand for superior goods, such as automobiles. A greater demand for superior or even luxury items has translated into greater global demands for petroleum. Circumstances not entirely anticipated by policymakers have increased the prominence of alternative energy on the policy agenda.

Policy Formation

The second step in the policy process involves the formation of policy solutions. Policy analysis occurs at this stage in the policy process; theoretically, a series of well-considered policy choices are studied and, through political and administrative processes, policy solutions are chosen for adoption. At the very least, successful policy formation requires that there is a generally accepted understanding of: (1) a policy problem; (2) the target population—individuals or things to be directly benefited or whose circumstances are to be altered by public policy solutions—and; (3) a clear sense of the costs and benefits associated with different policy solutions (Simon 2007). A solid understanding of a policy problem is often a function of social and physical science theory and findings regarding social or physical phenomena. In the case of alternative energy, understanding the policy problem requires an assessment of social values regarding energy—reduced air emissions—as well as future demands on energy—for example, prospects for the direction and growth of national and global economies. The target populations of energy policy range from the individual consumer and his or her energy demands to large-scale industrial users and energy producers to ecosystem needs. Understanding the target population(s) also requires an understanding of how policy will impact behaviors, conditions, and choices made by individuals and things whose conditions and behaviors are to be shaped by public policy. In essence, thorough policy formulators understand potential impacts of policy through the predictive aspect of social and physical science theories. Finally, direct and indirect costs and benefits associated with policy alternatives must be revealed in the policy-formation process so that choices between policy alternatives can be made and benefits can be maximized in relation to costs. Costs and benefits may be measured in economic terms as well as social and ecological terms, particularly in the case of alternative energy policy.

The unfortunate reality is that policy formation is often not successful in creating elegantly crafted and complete policy choices. Frequently, social and sometimes physical science theories are contested; target populations are not clearly understood or defined; and costs and benefits are insufficiently measured. As Deborah Stone (1988) notes, the numerical analyses so critical to good policy formation are often politicized—in essence, numbers are assigned meaning by policy analysts, a function of a preferred set of political and social values. A good example of this dilemma can be seen in the policy debate related to the Kyoto Protocol, where there is often noteworthy disagreement over the need or the potential impact of limiting greenhouse gas emissions.

Policy Implementation

Public policy often reaches the policy agenda because there is a perceived or real crisis which seemingly demands immediate attention. As John Kingdon noted in *Agendas, Alternatives, and Public Policy* (2003), opportunities to establish public policy are often short-lived; the policy "window" of opportunity may open and close quite rapidly, due to the short attention span of the public or policymakers. Thus, policy formation and subsequent implementation may occur at a fevered pace. In *Implementation* (1973), Jeffrey Pressman and Aaron Wildavsky catalog the challenging aspects of implementing public policy in an environment where it is expected that rapidly realized and successful policy impacts will emerge. Pressman and Wildavsky's study of the Oakland Project, a multi-million-dollar plan to revitalize the social and economic atmosphere of the California city, found that policy formulation and leadership were critical aspects of successful policy implementation.

Implementation of ill-considered plans that carry with them high hopes of success are, from the prospective of policy scientists such as Pressman and Wildavsky, unlikely. In the case of alternative energy, similar disappointments arose in the 1970s, as social and physical scientists became cognizant of the competing values and perspectives within their fields of study as well as public demand for the rapid implementation of an ameliorative solution to the energy crisis. Alternative energy policy analysts and implementers [(particularly advocates of Reagan energy policy approaches) were severely criticized] in the 1980s for a lack of effective response and called for significant reductions in alternative energy budgets (see Ahari 1987: 589). Perceived or real implementation failures have been used to further buffer the claims of free-market conservatives who argue that petroleum-based energy policy is functioning and will continue to function quite well without significant government policy intervention.

Policy Evaluation

Policy evaluation focuses primary attention on the intended and unintended impacts, outcomes, and relative (in)efficiencies of policy as it has been implemented. Most visible policy evaluation occurs at the end of a *policy cycle*, which can be defined in at least two different ways. A policy cycle commonly comes to an end every fiscal year due to the annual nature of the budget process. In deciding how and where to fund public policy, elected leaders and administrative personnel require some understanding of the relative success or failure of public policy as it has been implemented during the previous twelve months. In essence, evaluation plays a role in future *appropriation* decisions. A policy cycle also comes to an end when a public policy *sunsets*, a point in time usually between two and five years when Congress reviews the continued need for public policy in its present or adjusted form—that is, *reauthorization* decisions. Policy evaluation conducted at the point of reauthorization is another form of end-of-policy-cycle study and review.

Policy evaluation is an empirical review of policy processes, outcomes, and impacts on a defined target population as well as other nontargeted individuals and phenomena. Evaluation can be either qualitative or quantitative. Qualitative evaluations tend to focus significant attention on those aspects of public policy that are not easily measured in a uniform manner. Qualitative studies are often written in narrative form and may involve archival research or personal interviews with policy stakeholders—individuals or groups considering themselves interested in or impacted by a particular public policy. Quantitative studies are usually highly statistical, using numbers to represent a measured analysis of the processes, outcomes, and impacts of public policy.

An advantage to qualitative analysis is that there are many important, yet difficult to measure, aspects of policy studies. Absent qualitative analysis, many important aspects of policy outcomes, processes, and impacts would go unnoticed. A disadvantage to qualitative analysis is cost. It is time consuming to develop qualitative studies and it is equally time consuming to read and digest study results. Also, the validity and reliability of qualitative study is often difficult to determine. Quantitative study is able to measure the validity and reliability of findings produced in evaluative reports and summaries. Additionally, quantitative study deals directly in the language of policymakers concerned about budgets and reauthorization: namely, how much benefit was obtained for a given cost. A disadvantage to quantitative evaluation is that its very elegance and simplicity may not effectively underscore the impacts or outcomes of public policy on target populations and other stakeholders.

Policy Termination/Change

As noted above, public policies often *sunset*, requiring *reauthorization* every few years. During these periods of reauthorization, Congress and the president can make significant changes to the direction of public policy. The recent reauthorization of the Energy Policy Act (2005) is a good example of presidential and congressional reshaping of the priorities related to national energy policy. Additionally, public policy funding is generally decided on an annual basis. In both instances, it is in the power of elected officials — Congress and the president — to make decisions about the general direction of public policy. Agencies, which implement public policy, require money to operate. The power to shape an agency's budget, therefore, is one of the most powerful tools available to elected officials in shaping policy change on an annual basis. Congress also has the power of *legislative oversight*, which means that agency administrators can be called before congressional committees to discuss how policy is being implemented and to receive guidance on continuation or readjustment of policy direction (Aberbach 1990). At all of these points, policy can be changed in small and large ways. The courts should not be left out of this discussion, particularly in the case of policy change and/or termination. The courts can have a tremendous impact on public policy implementation and can, through common law decisions, impact the direction and goals of public policy. If a public policy is unconstitutional, the courts can effectively terminate it.

POLICY TYPES

In his 1972 *Public Administration Review* article, "Four Systems of Politics, Policy, and Choice," Theodore Lowi developed a typology of public policy. Central to his argument is the notion that government policy is essentially a forum in which individuals and groups make choices that they might not normally make without the presence of public policy. Lowi argued that "policies determine politics" (1972: 299). In other words, policy type impacts the political choices made. Sometimes, shaping choices involves coercion. In other instances, coercion is neither required nor practical. Through two-dimensional analysis, Lowi studies both the probability of coercion being employed (the vertical dimension) as well as the manner in which the coercion would be focused (the horizontal dimension).

According to Lowi, coercion can best be determined through legislative intent as evidenced by statute. In the table reprinted from his now classic journal article, Lowi argued that there were four types of public policy: constituent, redistributive, distributive, and regulative (see figure 2.2).

Constituent policy has a very low likelihood of being coercive and focuses primarily on shaping the environment in which conduct occurs (conduct is the outcome of choice). The establishment of the U.S. Department of Energy (DOE) is one example of constituent policy. President Carter sought to establish himself as a leader in environmental- and energy-related policy at a time when energy supply was of interest to the voters.

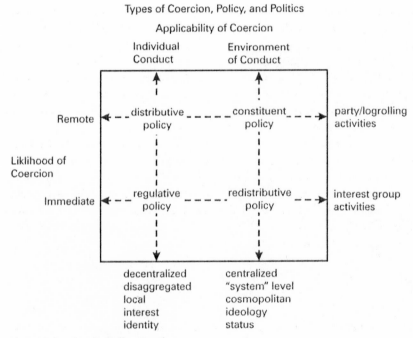

Figure 2.2.   Lowi's Policy Typology

Redisributive politics are similarly focused on environment of conduct, but frequently use coercion. Redistributive policy involves taking resources from one group and giving those resources to other groups who have less in the way of resources. In terms of energy policy, one form of redistributive policy might be energy tax credits for the elderly, which impacts the general environment in which conduct/choices occur.

Distributive policy focuses on decentralized efforts to shape individual conduct/choices, with very little likelihood of coercion. As an example, Lowi offers the nineteenth-century land grant program, through which significant portions of public lands in the West were deeded to individual citizens for their private use. The politics surrounding distributive policy tend to center on political parties, with some political parties more likely to represent a distributive policy area and to shape the manner in which that policy is formulated and implemented. Rural electrification is one example of distributive energy policy and is often closely associated with Democratic president Franklin D. Roosevelt, although rural energy issues have increasingly become the focus of rural Republican constituents and elected representatives.

A fourth policy type, *regulatory policy*, tends to focus on coercive methods of shaping conduct/choice while focusing to a greater extent on individual conduct. Individual conduct may involve the actions of an individual or corporate body in a particular individual instance. For example, if General Motors produced automobiles that violated U.S. pollution standards, then the corporation would be considered an individual violator. Regulatory policy impacts individuals or groups in particular instances. Interest in shaping the standards of regulation tends to fall on particular interest groups rather than political parties.

When energy "crises" occur, rapid policy change is often demanded. Immediate policy responses will emphasize the use of *regulatory* and *redistributive* policy solutions. Slower or more incremental approaches might emphasize *constituent* or *distributive* policy solutions. Politics shaping energy policy solutions will have both individual-focused as well as more environmentally focused aspects.

BOTTOM-UP POLICYMAKING/
TOP-DOWN POLICYMAKING

Policymaking can occur in different ways, but one fairly simple way to think about policymaking is to consider where policy ideas are first formulated and implemented (Dye 2001). *Bottom-up* policymaking focuses on policy ideas that are formulated and implemented at the local or state level of government. Bottom-up policy is often a function of public-private partnerships. Local-local or state-local intergovernmental partnerships and coordination are also commonly found in bottom-up policy efforts. Bottom-up policymaking was promoted in the 1980s and 1990s by governors and local leaders. A former state governor, President Clinton came into office in 1993, finding tremendous value in the "reinventing government" movement and the bottom-up domestic policy movement of that period. Bottom-up policy has the benefit of being implemented on a small scale in states or local areas, where policy experimentation may cost less, but outcomes could be highly innovate and effective. Successful policy experiments or innovations could then be implemented at the national level, possibly producing even greater successes (Osborne 1990).

Top-down policy approaches are less concerned with policy experimentation and are often more generally focused on protecting the rights and liberties of all citizens.[4] Social welfare, public health, environmental protection pollution standards, and civil rights policy areas could be considered examples of top-down policy, focused to a greater extent on national policy standards impacting all citizens, regardless of their location in the country.

Energy policy has examples of both top-down and bottom-up policy. Carbon-based energy policy is a well-established policy area and is regulated in a top-down policy approach. The DOE monitors and regulates the production, processing, transportation, and use of carbon-based fossil fuels (see Offices and Facilities, www.energy.gov, accessed November 25, 2005). Additionally, the Environmental Protection Agency (EPA) regulates the impact of carbon-based fuels on the environment. DOE and EPA policies are applied from the national level and impact individuals and corporations, regardless of location within the country.

Bottom-up policies have had and will continue to play a significant role in alternative energy policy developments. Various states and local governments have supported the development of alternative energy technology for electrical energy production. Photovoltaic panels have been used by schools and government buildings to cut the costs of government, saving citizens significant amounts of money that would normally go to lighting or operating government services (Wolfcale 2005). In Nevada, geothermal power will soon be used to produce hydrogen, a form of stored energy. Additionally, geothermal energy is used extensively to heat water tanks to develop fish farming in the desert Southwest.

COLLABORATIVE POLICYMAKING

Collaborative policymaking occurs on at least two levels. Within the bottom-up policymaking arena, the costs of policy innovation may exceed the capacity of any single governmental unit's budget. In order to produce tangible policy innovations at the local and state levels, government units *collaborate* to produce mutual benefits.

Collaboration is particularly important in the development of alternative energy. Government agencies, private sector businesses, and citizens often identify mutual benefits. Private industry benefits from the commitment of government to alternative energy and energy related products. Government and citizens gain the benefits of clean sustainable energy source development. California's solar roofs initiative is a good example of the mutual benefits emergent from collaborative policy efforts (see Wasserman 2003).

As is the case with other products and services, alternative energy gains legitimacy through use. Commitment to a goal and the innovation required to meet a goal on the part of government and the private sector are important parts of altering citizen consumption patterns. Many citizens are unaware of the energy-producing potential that exists in backyards and rooftops. Citizens are unlikely to invest heavily in the search process required to determine costs and benefits associated with alternative energy. Citizens may also be concerned with the aesthetic costs that alternative energy might produce (e.g., noise from wind generators, glare off solar panels, and panoramic views distorted by the physical plant of alternative energy generation). The NIMBY effect ("not in my backyard") emerges when alternative energy plans move from theory to practical application. Collaborative policymaking may help citizens become more aware of the costs and benefits expected to emerge from alternative energy policies. Sustainable energy paradigms for sustainable communities can only work if citizens are active participants.

CONCLUSION

In order to understand and accept a public policy, one has to get a sense of why the policy should even be in existence. A central issue in public policy is whether the policy is truly *public*. If a policy manages public goods, then it is very difficult to dispute a role for government in shaping the allocation of those goods or attempting to regulate quality and quantity. Air is quite obviously a public good, as was defined in this chapter. However, many other goods are not so easily placed into the public good category. Energy, for example, has often been seen as a marketable private good. Convincing individuals and groups that energy, particularly alternative energy, *ought* to be considered a public good is a bit of a challenge. It is, however, important to consider the role that energy will play now and in the future. Is it a public good? Are there certain public and private good aspects to the issue of energy, its production, use, and availability?

The public policy process, policy types, and the issue of collaborative policymaking are discussed because it is important to understand how policy comes to fruition and what possible foibles might befall public policy as it moves through the process. Policy can be thought of from a variety of different angles; Professor Lowi does provide some food for thought in that regard. As you might notice from the discussion of policy types, different types of policy produce different types of politics. Alternative energy policy seems to be heading predominantly toward interest group politics, which may not advance individual goals in this area in a uniform manner. Collaborative policymaking is an inviting way to overcome some of these dilemmas. Bottom-up policy innovations are particularly important to the future of alternative energy and sustainable communities.

NOTES

1. See Rochlin (2004). Rochlin (p. 52) uses the California energy crisis as an example of a similar mindset. As Rochlin points out, increases in energy prices were "viewed as part of the problem...not a symptom, let alone a solution, to an imbalance in supply and demand." If energy reserves are considered a public good by government, then it is crucial that government accurately predict future demand sans market mechanism. Houldon (2004: 66) counters Rochlin's argument, concluding, in essence, that energy is a central aspect of modern society and has become part of the social contract related to individual rights. Houldon states, "the true public good is the energy itself, which cannot be separated from any other aspect of the infrastructure without incurring significant transaction costs nor from any of its attributes, such as security and the reserves that are a part of the means of providing security."

2. In chapter 1 of Simon (2007: 1), public policy is defined as what "government ought or ought not do, and does or does not do."

3. Of course, the vice president is nationally elected, but usually not an overtly active agenda setter. It has been noted that Vice President Dick Cheney has played a historically atypical role in shaping public policy at the national level (see Krugman 2004).

4. At times, top-down approaches are focused to a greater extent on policy elites' priorities rather than majoritarian benefits.

WORKS CITED

Aberbach, J. 1990. *Keeping a Watchful Eye: The Politics of Congressional Oversight.* Washington, D.C.: Brookings Institution.

Aberbach, J. and Rockman, B. 1978. Administrators' Beliefs about the Role of the Public: The Case of American Federal Executives. *Western Political Quarterly* 31(4): 502–22.

Ahari, M. 1987. Congress, Public Opinion, and Synfuels Policy. *Political Science Quarterly* 102(4): 589–606.

Binder, S., Lawrence, E., and Maltzman, F. 1999. Uncovering the Hidden Party Effect. *Journal of Politics* 61(3): 815–31.

Bowers, B. 2004. Stone Age Combustion. *Science News* 165(18): 276–77.

Cook, D. 2003. Spencer Abraham. *Christian Science Monitor*, September 12: 25.

Dye, T. 2001. *Top Down Policymaking.* New York: Chatham House.

Hayek, F. 1944. *The Road to Serfdom.* Chicago: University of Chicago Press.

Houldon, R. 2004. Finding the Public Good: Shedding Light on a Bulk Grid Electricity Card Trick. *Electricity Journal* 17(9): 61–67.

Hulbary, W. and Walker, T. 1980. The Supreme Court Selection Process: Presidential Motivations and Judicial Performance. *Western Political Science Quarterly* 33(2): 185–96.

Jones, Charles O. 1970. *An Introduction to the Study of Public Policy.* Belmont, CA: Wadsworth.

Kingdon, John W. 2003. *Agendas, Alternatives, and Public Policy*, Third Edition. New York: Longman.

Krehbiel, K. 1993. Where's the Party? *British Journal of Political Science* 23(2): 235–66.

Krugman, P. 2004. A Vision of Power. *New York Times*, April 27, A25.

Lowi, T. 1972. Four Systems of Politics, Policy, and Choice. *Public Administration Review* 32(4): 298–310.

Michelbach, P., Scott, J., Matland, R. and Bornstein, B. 2003. Doing Rawls Justice: An Experimental Study of Income Distribution Norms. *American Journal of Political Science* 47(3): 524–39.

Nation's Cities Weekly. 2000. Bush and Gore Address Environmental, Energy Issues. *Nation's Cities Weekly*, October 23: 16.

Osborne, D. 1990. *Laboratories of Democracy.* Boston: Harvard Business School.

Ostrom, E. 1990. *Governing the Commons: The Evolution of Institutions for Collective Action.* New York: Cambridge.

Pressman, J. and Wildavsky, A. 1973. *Implementation: How Great Expectations in Washington Are Dashed in Oakland; Or, Why It's Amazing that Federal Programs Work at All, This Being a Saga of the Economic Development Administration as Told by Two Sympathetic Observers Who Seek to Build Morals on a Foundation of Ruined Hopes.* Berkeley: University of California Press.

Rawls, J. 1971. *A Theory of Justice.* Harvard, MA: Belknap Press.

Rochlin, C. 2004. Resource Adequacy Requirement, Reserve Margin, and the Public Goods Argument. *Electricity Journal* 17(3): 52–59.

Simon, C. 2007. *Public Policy: Preferences and Outcomes*. New York: Longman.

Stone, Deborah A. 1988. *Policy Paradox and Political Reason*. New York: Harper Collins.

Unger, R. 2004. *Mills in the Medieval Economy: England 1300–1540*. New York: Oxford University Press.

Wasserman, J. 2003. Schwarzeneggar's Growth Agenda May Rile State's Suburban Builders. *Associated Press*, November 9, State & Regional section.

Weatherford, S. 1988. The International Economy as a Constraint on U.S. Macroeconomic Policymaking. *International Organization* 42(4): 605–37.

Wilson, J. 1989. *Bureaucracy: What Government Agencies Do and Why They Do It*. New York: Basic Books.

Wolfcale, J. 2005. Going Solar in Kentfield. *Marin Independent Journal*, September 28, Local News.

WEB SITE

Offices and Facilities, www.energy.gov, accessed November 25, 2005.

Chapter Three

Overview of Alternative Energy and Fuels and Their Uses

INTRODUCTION

Previous chapters outlined the trends that have led the United States in the direction of alternative energy. Social and political events of the 1960s and 1970s were responsible for the initial shifts in the energy paradigm. The book has also introduced the reader to ways of thinking about public policy, particularly in relation to alternative energy.

In this chapter, the book focuses on the scope of *alternative energy*, a concept that is often discussed but rarely defined. It is difficult to define because the term is value laden. The term "alternative energy" was the subject of a recent gathering of scientists in Canada, but by the end of the conference, the definition remained a work in progress; no definitive meaning was assigned (see CEA-NRCan 2002). By defining the term, it is not my purpose to intentionally exclude or include any particular form of energy. An understanding of the concept is necessary to better understand the energy future in relation to technological, economic, and policy feasibility of energy sources. Understanding the difference between alternative energy and alternative fuels is an important distinction that must be made at this point.

WHAT IS ALTERNATIVE ENERGY?

The federal definition of alternative energy is best summarized by Title 26, chapter 79, §7701 of the revised U.S. Code: "the term 'alternative energy facility' means a facility for producing electrical or thermal energy if the primary energy source for the facility is not oil, natural gas, coal, or nuclear power." The primary purpose of this definition relates to the issuance of tax credits to "alternative energy facility[ies]," which meet certain standards as defined in Title 26, chapter 1, §48 "Energy Credit." Tax credits are one method by which the federal government encourages the private sector to make certain economic choices; in the case of energy policy, this definition of alternative energy will have a definitive impact on how alternative energy will be defined by those individuals and corporate bodies seeking federal recognition (and benefit) by adopting a particular definition of alternative energy. Many state definitions of alternative energy closely follow federal definitions. Case law confirms that federal guidelines supercede state-level guidelines. Federal standards also impact the state and local receipt of alternative energy grants, subsidies, and tax exemptions. It is reasonable, therefore, that state and local definitions would be consistent with federal energy policy. Consistency between federal and state definitions does not mean that there are not a few variations. In many ways, variation at the state level illustrates the dynamic and evolving alternative energy paradigm, which is by no means unique to the U.S. policy process.

Energy versus Power

These terms are often used interchangeably and are thought to be the same concept. Energy is defined as "the capacity to do work. Forms of energy include thermal, mechanical, electrical, and chemical. Energy may be transformed from one form into another" (EERE 2005a). Energy can be measured in a variety of ways, such as joules and British thermal units (BTUs). Power is "the rate of expenditure of energy" (Physics Forum 2005) or the rate at which energy is used. Power is simply expenditure of energy per unit of time (e.g., BTU/hr or BTU/sec). In everyday life, one comes across measurements of power in the form of watts or horsepower. Fuel is potential energy that can be burned to produce power.

State-level variation illustrates the role politics plays in shaping the alternative energy paradigm. State-energy economic interests often impact the choices related to the adoption of alternative energy policies. For example, in states with coal production, it is likely that interest groups representing fossil fuel industries and environmental interests will square off, seeking to shape the regulation of energy uses and production as well as the nature of redistributive policies intended to provide cleaner energy, the latter effectively benefiting individuals and groups who may bear environmental and health costs associated with fossil fuel use. In some cases, there are even different tiers or categories of "alternative energy," depending on the source's level or shades of "greenness" (Mandelbaum and Brown 2004: 1).

Perhaps the most well-known recent example of very broadly defined and apparently "relaxed" standards for the concept "alternative energy" comes from the state of Pennsylvania, which has effectively moved away from more narrowly defined federal definitions. Adopted into law in November 2004, Pennsylvania Senate Bill 1030 identifies energy sources and relates these sources to relative "greenness," which is by implication measured by the level of environmental emissions or potential for environmental harm. The alternative energy greenness divides types into Tier I and Tier II alternative energy sources.

Tier I alternative energy sources are: (1) solar PV energy; (2) wind power; (3) low-impact hydropower;[1] (4) geothermal energy;[2] (5) biologically derived methane gas;[3] (6) fuel cells;[4] (7) biomass energy;[3] and (8) coal mine methane. In scrutinizing these Tier I sources and the related definitions in the footnotes, the reader might dispute their categorization as relatively low impact forms of energy. Coal mine methane and biomass derived energy, for instance, produce greenhouse gases. Pennsylvania's Tier II alternative energy resources provide an even broader view when it comes to common associations with "green" energy. The second-tier sources are as follows: (1) waste coal; (2) distributed generation systems; (3) demand-side management; (4) large-scale hydropower; (5) municipal solid waste; (6) generation of electricity utilizing byproducts of the pulping process and wood-manufacturing process, including bark, wood chips, sawdust, and lignin in spent pulping liquors; and (7) integrated combined coal gasification technology.

The Pennsylvania statute illustrates the capacity of government to narrowly or broadly define alternative energy. Alternative energy cannot be assumed to have certain properties, such as zero greenhouse gas emissions or complete disassociation from hydrocarbon energy sources. At the federal level, the expansion of nuclear energy policy has been proposed as a form of alternative energy, replacing the use of hydrocarbons to produce electrical or thermal energy. What is alternative energy? Simply put, it is not a unified concept. The lack of clarity and consistency in definition, however, provides an opportunity for individuals and groups considering alternative energy sources for their communities, states, or nation. The lack of rigidity in definition reflects the likelihood that perspectives on alternative energy will be more likely related to a search process focusing on energy alternatives.

WHAT ARE ALTERNATIVE FUELS?

Federal law is fairly definitive when it comes to alternative fuels. Alternative fuel issues are often tied directly to their primary use: transportation. Federal fuels policies bring together issues related to stored energy sources and management, transportation infrastructure, and environmental quality regulations as well as federal monies to state and local governments. The federal definition of alternative fuel is found in Title 42, chapter 77 §6374 of the U.S. Code:

> The term "alternative fuel" means methanol, denatured ethanol, and other alcohols; mixtures containing 85 percent or more (or such other percentage, but not less than 70 percent, as determined by the Secretary, by rule, to provide for requirements relating to cold start, safety, or vehicle functions) by volume of methanol, denatured ethanol, and other alcohols with gasoline or other fuels; natural gas; liquefied petroleum gas; hydrogen; coal-derived liquid fuels; fuels (other than alcohol) derived from biological materials; electricity (including electricity from solar energy); and any other fuel the Secretary determines, by rule, is substantially not petroleum and would yield substantial energy security benefits and substantial environmental benefits.

ALTERNATIVE ENERGY SYSTEMS: OVERVIEW OF HOW THEY WORK AND POTENTIAL USES

It would seem fairly self-evident that energy sources could be used to generate electricity and that electricity could be used for vehicle transportation purposes. Basic technicalities aside, potential use of an alternative energy system is predominantly a function of economic incentives and regulations. In the case of alternative energy, the costs associated with wind, solar, and geothermal energy have declined, particularly in relation to the direct costs of fossil energy sources, making such energy sources economically feasible alternatives. Additionally, energy regulations, such as PURPA (Public Utility Regulatory Policies Act of 1978), have increased the demand for alternative energy by requiring that "utilities . . . purchase power from small power projects that [use] renewable sources of energy" (Bettelheim 2000: 8).

Finally, grants and tax incentives, and the possibility of "net metering," which allows alternative energy producers to sell green power for use on the power grid, have made it more likely that businesses, local governments, and individuals will develop alternative energy. Alternative fuels policies primarily employ vehicle emissions standards to regulate the automobile industry, while agricultural subsidies, a form of redistributive policy, are used to encourage farmers to grow corn for use in the production of ethanol. Taking a broader definition of "alternative energy," producer subsidies and tax incentive structures are used to encourage fossil fuel processing to explore gas fields for methane and propane, as well as more efficiently capture and use of coal field gases that would otherwise be released into the atmosphere—misallocated energy resources.

Although social and political values, as highlighted in chapter 1, play a role in the use of alternative energy, economic values are extremely important. The incentive structure established through regulation is designed to meet social and political values and is also intended to increase supply of alternative energy. From an economic perspective, increased supply may actually lower direct cost to consumers on a per unit price basis for particular goods or services.

Alternative energy is currently at a very critical point in which market forces are more inclined to use fossil energy because of cost factors, yet those fossil fuel costs are rising because of increased demand—and the production, storage, and transportation infrastructure associated with fossil energy supply is struggling to meet demand. The visibility of alternative energy as a resource is being highlighted through government grants and tax incentives, as well as through constraints on fossil-energy processing (e.g., constrained petroleum refinery capacity).[5] With increased constraints on fossil energy and a positive incentive structure developing for alternative energy, it is important to develop a general understanding of the capability of alternative energy sources.

Solar Energy

Most individuals are familiar with solar energy at some basic level. After the first sunburn on the ski slopes, however, people know that even when it is cold outside, the sun radiates energy. So, whether consciously aware of it or not, most individuals figure out that the sun's energy has a wide variety of effects and uses. The trick comes in capturing that energy and transforming it into electricity.

Photovoltaic (PV) cells are used to capture solar energy and convert it into electricity. A PV cell is a semiconductor, which means that the current carrier (electrons) in the material is influenced by the input of energy. The introduction of energy excites electrons in semiconductors and these electrons may form a stream (i.e., electrical current). In a PV cell, electrons in silicon semiconductors are impacted by the introduction of solar energy and begin to freely move about. Metal strips embedded in the silicon PV panels collect these free-flowing electrons, which then move in a current and are collected from the panel. Solar thermal devices are not directly concerned with producing electricity. Instead, solar thermal systems concentrate solar energy so as to produce heat "at useful temperatures" (www.eia.doe.gov/cneaf/solar.renewables/page/solarthermal/solarthermal.html, accessed May 31, 2005). Solar thermal systems and solar PV systems come in various sizes and can be used for both home and commercial purposes. In the case of solar PV systems, the energy produced is direct current (DC) that is often stored in batteries (also DC). Commonly used household appliances in the United Stated demand alternating current or AC, which requires that DC be converted to AC before use.

Safety, Environmental Damage, and Emission-Related Issues: Solar Energy

In the case of standard PV systems, user safety issues are minimal. As one source noted, the biggest danger is falling off of the roof during home installation or during the maintenance process (Wholesale Solar Electric 2005 [www.wholesalesolar.com/products.folder/mount-folder/mount-info.html, accessed November 26, 2005]). Most safety issues relate to the material used in the production of solar panels. For example, tin oxide (SnO_2) used in electron collector grids is a potential soil and water contaminant.

Solar energy systems that require concrete footings to support collectors are potentially environmentally damaging to the degree that vehicle operation and land clearing might disturb native plant and animal species. It is conceivable that water and soil contamination could occur as a result of system installation. Many solar systems require cleaning to prevent light refraction that reduces the efficiency of solar collectors. Dust and other forms of dirt may collect on solar energy collectors and reduce the ability of these collectors to be properly exposed to the sun's rays. Water or industrial solvent cleaners may be needed to clean solar collectors. Although water is not damaging to the environment, the use of water may have an impact on groundwater and native streams, and may impinge on other animal or plant species use of the resource. Industrial solvents may be harmful to groundwater and may cause soil contamination in high concentrations.

The Price of Success: Inside the New Jersey Clean Energy Program

How the second largest solar program in the United States is dealing with the ups and downs of being an industry leader, by Stephen Lacey, RenewableEnergyAccess .com

Trenton, New Jersey [RenewableEnergyAccess.com]. The overwhelming response to the New Jersey Clean Energy Program (CEP) rebate system has contributed to significant delays for commercial and residential rebate approvals—hurting many solar companies that are supposed to benefit from the program.

The purpose was to create rampant growth in the solar industry. But in retrospect, it happened too well. That has caused a situation where companies in the industry are feeling very desperate because new commitments are not being made; therefore, businesses have stagnated. They are desperately hanging on, most are laying off people, and some may have to go out of business. (Lyle Rawlings, New Jersey Clean Energy Council member)

The issue has brought solar professionals, government officials and financiers together to craft a workable solution that will free up delayed projects and get the CEP—one of the best incentive based programs in the country—running smoothly again. But some in the industry say the New Jersey Board of Public Utilities (BPU) has not acted quickly or decisively enough to mitigate the financial blow to solar companies.

The rebates, which formerly covered up to 60 percent of an installed solar system, had been a great success. And as a result, there are now many applications sitting in a queue. Both the rebate levels and consumer demand have been very high. In an effort to temper the program and limit applications, rebates have been lowered 5 times in 15 months.

Also, to meet the strong solar goals of New Jersey's Renewable Portfolio Standard (RPS), the BPU made a large number of commitments to solar projects. However, those commitments exceeded the CEP budget through 2008, putting a 9-month halt on up to 200 projects.

"The purpose was to create rampant growth in the solar industry, but in retrospect, it happened too well," said Lyle Rawlings, a member of the New Jersey Clean Energy Council. "That has caused a situation where companies in the industry are feeling very desperate because new commitments are not being made; therefore, businesses have stagnated. They are desperately hanging on, most are laying off people, and some may have to go out of business."

Rawlings, who is also President of the New Jersey-based company Advanced Solar Products, said that growth of his business has "essentially stopped" because of the long queue for applications at the BPU.

For a while, things were looking good for solar companies. But then assurances for materials started to fall through. Businesses were told by the BPU that they could order solar modules for projects within months, but when applications were stalled indefinitely because of budget constraints, companies had to cease operations. Consequently, it became difficult to plan for projects, and brought the industry to a standstill.

"From December 22, 2005 to September 22 of 2006, the program was in a limbo," said Richard King, President of American Energy Technologies (AET). "There was this cascade effect within the community for electricians, roofers, crane operators, delivery drivers, distributors and dealers who were just ramping up the program."

According to King, AET, a renewable energy installation company based in New Jersey, currently has tens of millions of dollars in projects waiting in line at the BPU. It could be anywhere from 9 to 16 months before any of the projects start moving again.

While the delay will be difficult for many solar companies in the state, the CEP has a different view. New Jersey's RPS, which set a goal of 1,500 megawatts (MW) of installed solar by 2020, requires aggressive action. By setting high rebates and

(continued)

allowing a steady flow of applications, the state has seen installations of roughly 22 MW of solar in the last 5 years. That, said CEP Director Mike Winka, has been very good for solar businesses.

"We have a very vibrant program," said Winka. "We are working very hard to meet the solar requirements set by the RPS. And with a proper switch from rebates to a Solar Renewable Energy Certificate [SREC] financing system, the program will be that much better."

AET's King agreed with that assessment. However, while he praised the intentions of the program and faulted no one in particular, he found it "unforgivable" that companies have suffered so badly. He felt that the BPU has not acted fast enough to transform the program and get a full SREC system in place as promised.

An SREC system is a market-based payback option supported by clean energy generation Alternative Compliance Payment (ACP) options, which are paid by the electric utilities within their operating expenses. Owners of solar arrays obtain an SREC each time they generate 1 megawatt-hour of electricity.

Currently, one SREC is worth about $200. The owner trades those certificates on the Clean Power Market, allowing utilities to buy the clean power credits. Eventually, if the value of SRECs can double, that will allow the CEP to phase out the rebate program. But it will take time to make a transition from rebates to SRECs.

"We're going to move as quickly as possible to put this system into place," said Winka. "But we can't move too fast. We need to be cognizant of the regulatory structures. The financial industry needs to have trust in those regulations."

There are many differing views on how the new SREC system should work. The Office of Clean Energy has posted White Papers on its website authored by members of the Clean Energy Council. The papers offer ways to transition from the rebate program to an SREC financing system. The CEP already offers SRECs as an option to consumers. But in June of 2007, a broader 17 MW pilot program will be put in place to gauge how best to implement a full certificate system.

The problems facing the New Jersey CEP are paradoxical. Businesses are hurting because of delays; however, the delays have been caused by the growing demand for solar. Now it is a matter of managing the program's success so that everyone will be satisfied.

Indeed, sometimes being a leader means making mistakes first—then coming up with the best solutions. Other states will be watching closely to see how the New Jersey CEP makes a transition from rebates to SRECs. The outcome will certainly provide some valuable lessons for all.

(www.renewableenergyaccess.com/rea/news/story;jsessionid=6DFE9B5975406FAB64C77143467B CE6B?id=46172, accessed October 12, 2006).

Wind Energy

While PV systems are relatively new, wind power is probably as old as the wheel and the development of agriculture-based communities. Wind energy requires the rotary motion of propeller type blades to collect energy. The first vertical axis windmill was developed in Persia (c. 500–900 B.C.), was used to operate grist mills for processing grain crops (telosnet.com/wind/early .html, accessed May 31, 2005). Other scholars claim that the windmill was first developed and used in China over two millennia ago (telosnet.com/ wind/early.html, accessed May 31, 2005]). In Western Europe, early windmills were erected on a single timber and were known as "postmills." In the late fourteenth century A.D., the Dutch developed larger windmill systems mounted atop towers. Windmills used for mechanical water pumping were perfected for small-scale use in the United States during the nineteenth century.

An electricity-generating windmill was first developed in 1888 in Cleveland, Ohio. Named after its creator Charles F. Brush, the Brush machine was capable of producing twelve kW of electricity (note: a clothes dryer requires five kW; see www.psnh.com/Residential/ReduceBill/Applianceusage.asp, accessed May 31, 2005). During the early twentieth century, wind power was used on farms in the United States. Electricity demands were growing and the cost and availability of private commercially generated power were beyond the reach of many potential consumers, particularly individuals living in rural areas. The Rural Electrification Act of 1936 (7 U.S.C. 31) soon made available relatively cheap electricity to even remote rural areas and wind power systems declined in use as they became viewed as an inferior source of energy (www.usda.gov/rus/regs/info/100-1/title_i.htm, accessed May 31, 2005).

In the post–World War II period, technical and commercial developments in wind power technology occurred primarily in the Soviet Union and Europe. Germany and Denmark became leaders in technological development and commercialization. Wind turbine and blade technology advances have made wind systems more efficient in energy production. One of the most advanced European commercialization efforts has been conducted by the Vestas Corporation, which currently produces a variety of wind turbine technology up to 4.5 megawatt (MW) systems for land- or sea-based application (Vestas 2005 [www.vestas.com/uk/Products/products2004/prodOverview_ UK.htm, accessed November 26, 2005]).

It was not until the 1973 oil embargo that the United States began in earnest to pursue the development of modern wind technology systems. Nationally sponsored centers, such as Sandia National Laboratory, and public-private partnerships have played a significant role in developing wind turbine technology for the twenty-first century. Wind energy technology efficiently and effectively operates systems at various wind speeds and to reduce the costs associated with wind energy production. Sandia National Laboratory and its private-sector partners are currently developing systems that will eventually reduce low wind speed energy production costs to 3¢/kWh (the number of kW generated on an hourly basis) (National Wind Speed Technology Center 2005 [www.nrel.gov/wind/about_lowspeed.html, accessed November 26, 2005]).

Safety, Environmental Damage, and Emission-Related Issues: Wind Energy

Environmental damage from wind energy is increasingly understood as wind projects begin to expand across the country. At least two major issues are often discussed regarding the impact of wind energy systems on the environment:

- noise pollution
- visual pollution

These issues are alternately rejected and supported. Agreement on the impacts of wind energy systems is hard to find. If one accepts Lowi's thesis that energy policy is both regulatory and redistributive policy, then it would be no surprise that different policy interests generate evidence to support their preferred policy views, while rejecting the views of other interests.

In small-acreage rural/urban fringe settings, even solitary wind energy systems may cause neighborhood noise. At times, community development regulations prevent the installation of wind generation systems, but in many rural/urban fringe areas this restriction may not be possible. Large-scale wind generation systems located in more remote areas create noise pollution that may negatively impact native animal species as well as nearby residents. It is known that noise pollution has a tendency to cause native animal species to relocate, which may negatively impact the cycle of nature in locations abandoned as well as new preferred locations. Animal species may find themselves constrained by noise pollution and nearby human residents, creating confrontations between animals and humans that result in lose-lose situations for all concerned.

Visual pollution is also of concern with wind system applications. As seen in figure 3.1, a large-scale wind farm can completely change the visual image of an otherwise bucolic scene. Visual pollution is, like beauty, in the eye of the beholder. For some individuals, the visual "cost" may not outweigh the individual and collective benefit of having a clean source of energy, such as wind (BBC News 2005 [news.bbc.co.uk/1/hi/sci/tech/4560139.stm, accessed November 26, 2005]). Local property owners often cry "visual pollution" when they think that a wind farm will negatively impact their property values.

Figure 3.1. Wind Turbines and Visual Pollution
Source: Available at www.sandia.gov/news-center/news-releases/2003/images/jpg/turbine-land.jpg, accessed June 10, 2006.

There has been some discussion of the possibility of climatological impacts of wind generation systems. Nonmainstream scientists and other concerned individuals speculate that large-scale systems—such as those found in northern California—may change the regional weather. At this point, mainstream scientists find little credence in the argument that weather patterns will be altered by wind farms. Climatologists point out that the impact of the use of fossil fuels and the gravitational influences of the moon on the Earth are more readily detectable than any possible impact of "wind farms," although it is conceivable that large-scale wind generation facilities could minutely slow the wind in the region (www.bbc.co.uk/norfolk/talk/wind_farms.shtml, accessed June 1, 2005).

In early wind system applications, avian mortality was a problem. Bird death was a function of the powerful and rapidly rotating propeller blades on modern wind generation systems. As the wind passes through turbine propeller blades, birds flying in close proximity were likely to become dragged into the inflow and killed. While avian mortality was a serious environmental impact, the fact that endangered native species of bird were more likely to be harmed in these more remote native settings was particularly troubling (see www.awea.org/pubs/factsheets/avianfs.pdf, accessed June 1, 2005).

Recommendations emergent from a recent study conducted in California are expected to significantly reduce avian mortality issues (see Smallwood and Thelander 2004).

Wind Energy: Middlebury College, Vermont

Middlebury College Receives $22,500 Grant for Wind Energy System
College to provide matching funds
Contact:
Adrianne Tucker
satucker@middlebury.edu
(802) 443-5629

Release Date: Wednesday, May 11, 2005

MIDDLEBURY, Vermont—Middlebury College has received a $22,500 grant from the United States Department of Energy to construct a wind turbine. The project is part of an initiative that will assist the State of Vermont as it explores the technology necessary for wind-generated electricity. The Vermont Department of Public Service (DPS) will administer the grant. The college will provide matching funds to complete the project, which will include collecting information on available wind resources, offering educational outreach, and assessing the value of net-metering for Vermont schools. Net-metering is a technology that allows small energy producers to feed their unused power back to their commercial suppliers for credit on their accounts.

"This grant will help the college make progress in accomplishing its carbon reduction goals," said Bob Huth, vice president of administration and treasurer of Middlebury College. "The educational opportunities related to wind energy that the grant creates will benefit both the college and the larger Vermont community. We're very encouraged by it."

Once the net-metering permit has been approved by the Vermont Public Service Board in June, the college will hire Vermont Green Energy Systems of East Middlebury to construct a wind turbine on the western edge of campus. The $45,000 wind energy project is slated to begin construction in mid-June. Upon completion in early September 2005, the turbine will be open to the public. Area schools will be invited to tour the turbine facility, and the data it collects will be available for schools to include in curricula on renewable energy.

"The installation of this wind turbine will be part of the 40th anniversary celebration of our environmental studies program," said Middlebury College Director of Environmental Affairs Nan Jenks-Jay. "It reflects the college's commitment not only to environmental education and research, but to its sustainable campus program as well."

Waitsfield-based Northern Power Systems, which designs, builds, and installs flexible onsite and integrated power systems, assisted with the development of Middlebury College's turbine project. To be located at the site of the campus's recycling

(continued)

facility, the turbine will be connected to the college grid, offsetting the college's use of electric power from Central Vermont Public Service. The Middlebury College Recycling Center will use as much of the wind-powered electricity as it needs for operation at any given time. Electricity not utilized by the center will be fed through the grid and used elsewhere on campus. According to Mike Moser, assistant director of the college's facilities management department, the proposed wind turbine will produce more than 15,000 kilowatt hours per year—roughly equivalent to the annual energy consumption of a home powered entirely by electricity.

Moser links the college's wind turbine project to its carbon reduction initiative. "Every kilowatt-hour generated by wind instead of fossil fuel prevents air pollution and greenhouse gas emissions," he said. "We need to be exploring this technology."

Middlebury College students participating in the wind energy project will provide guided visits to the turbine for local schools, and develop a Web page where data on its wind-generated electricity will be published and regularly updated. According to Amy Seidl, an associate in science instruction at the college, there is a growing interest among students in wind technology.

"My students have already begun to define research projects that will focus on seasonal and wind parameter differences in electricity generation, as well as service-learning investigations of the potential economic benefit for farmers in the Champlain Valley," she said.

The DPS has administered federal funds for other wind energy projects in the state, including wind mapping and measurement projects, a wind system for the state at the Alburg Welcome Center, and the Vermont Environmental Research Associates' meteorological tower on Lake Champlain.

Source: web.middlebury.edu/offices/pubaff/news_releases/news_2005/20050511wind.htm, accessed May 31, 2005.

GEOTHERMAL ENERGY

Geothermal energy uses the heat of the Earth to produce energy. The Earth's mantle is composed of superheated iron and other elements. In some locations, this superheated material rises into the Earth's crust and close to groundwater supplies. Deep wells are drilled in areas of accessible geothermal activity and the superheated water or steam is extracted. Since most geothermal waters have high salt content, some geothermal water is commonly called brine.[6] A closed system allows the superheated brine to transfer its energy to other low-boiling point materials. The low-boiling point liquids in the secondary stage system is used operate high-efficiency turbines, which in turn produce electrical energy. The water is then reinjected into the geothermal aquifer in close proximity to the extraction well.

Geothermal resources are graded in terms of the extracted water temperature and the relative size of the geothermal field (i.e., how much superheated water exists in a particular location and the depth of the well needed to efficiently extract the resource). According to the DOE, geothermal resources above 200° C are high-grade geothermal resources. Geothermal resource temperatures can be as high as 700° C. As you will likely notice, most of these high-grade resources are located in the American West. Medium-grade resources (150–200° C) are located in the West and Southwest, while low-grade resources (100–150° C) are found throughout the United States.

There are various types of low-boiling temperature fluids used in the heat exchange process. Typically, organic materials that are gases at room temperature, such as pentane or butane, are used in binary systems for the heat transfer. Pentane and butane have low boiling temperatures (36.1° and −1° C, respectively), which means that the transfer of energy from even a low-grade geothermal resource would have a tremendous effect on the state of these secondary products in the binary system heat transfer and turbine operation.

Safety, Environmental Damage, and Emission-Related Issues: Geothermal Energy

There is potential for geothermal energy resources to cause environmental damage and to produce airborne emissions, but problems can be minimized through the use of a closed-loop system. A closed-loop system withdraws, uses, and reinjects brine, maintaining it within the confines of piping during the surface process. No brine or brine steam is released into the air in a closed loop, which reduces the potential for contamination of ground water and surface air. Some of the contamination that may be released in an open-loop system are as follows (see www.ucsusa.org/clean_energy/health_and_environment/page.cfm?pageID=942, accessed June 2, 2005):

- hydrogen sulfide
- ammonia
- methane
- carbon dioxide
- toxic sludge containing sulfur, arsenic, mercury, nickel, etc.

Another potential issue related to geothermal resource use is the location of geothermal wells and equipment on public lands and environmentally sensitive areas. With proper management of the resource and its use, it might be possible to maintain a reasonably small technology "footprint" for geothermal resource extraction and use.

Geothermal Energy: Pinal County, Arizona

Arizonans Tap into Geothermal Energy

Some farms and communities in western Pinal County may be sitting on a valuable natural resource called geothermal energy, which could help heat homes and benefit farmers, experts say.

"Anything that you can think of where you need heat, you can do with geothermal," said Mike Pasqualetti, professor of geography at Arizona State University.

Geothermal means "earth heat" or heat from the earth's interior. There is a heat source beneath the surface everywhere in the world, Pasqualetti said, but it is closer to the surface and hotter in some places.

Volcanic eruptions, geysers and hot wells are signs of geothermal activity below the surface.

Those geothermal resources exist in this part of the state, including a corridor from Eloy and Picacho to Coolidge, and the U.S. Department of Energy wants to help develop them, Pasqualetti said.

The Department of Energy has awarded a grant to Northern Arizona University in collaboration with ASU, New Mexico State University, Arizona Public Service Co., and the Ormond Group to help farmers and others in Arizona investigate possible geothermal uses.

Some wells in the corridor have water temperatures ranging from 122 to 176 degrees Fahrenheit, Pasqualetti said. That is hot enough for a destination spa or to heat buildings and subdivisions.

The Oregon Institute of Technology's Geothermal Resource Database identifies eight hot wells within five miles of Coolidge.

Some wells around Arizona City and Maricopa have water temperatures that range from 95 to 113 degrees Fahrenheit, Pasqualetti said. That is hot enough to heat green houses and raise fish and shrimp.

Pasqualetti said people have used geothermal energy for thousands of years for bathing, medical therapy, cooking, even heating. Houses in the ancient Roman city of Pompeii were heated with geothermal water.

People use it today to heat homes and districts in Boise, Idaho; Klamath Falls, Ore.; Susanville, California; and San Bernardino, California, among other places.

A hotel in New Zealand uses it for heating and cooling.

Twenty-four countries generate electricity with geothermal resources. A geothermal power plant north of San Francisco generates enough electricity to power San Francisco and Oakland.

APS has received a Department of Energy grant to dig an exploratory "slim hole" in the Clifton area to see if there is potential for a geothermal power plant, said Amanda Ormond, CEO of the Ormond Group.

Geothermal energy operates all the time, Pasqualetti said, unlike solar collectors and wind turbines, which shut down when the sun sets or the wind stops blowing.

It is clean, unlike fossil fuels, and the water can be injected back into the aquifer after the heat is extracted.

(continued)

Geothermal power plants also do not have to be large to be economical, he said. So they can be built in areas that could never support a large nuclear-, coal- or gas-fueled power plant.

New Mexico has 50 acres of greenhouses heated with geothermal energy, according to the Department of Energy.

Elko, Nevada, has geothermal laundries, Pasqualetti said. Outside Reno, Nevada, is a geothermal onion dehydration facility that has been making money for 20 years.

"They have their heat source for free once they drill the well and amortize that cost," he said. "They put the whole facility right there, because that is where the resource is."

Ormond said geothermal resources could be a benefit to dairy farmers, for example, who heat a lot of water to sterilize milking equipment.

A few places in Arizona already use geothermal energy, Pasqualetti said. Shrimp are raised in geothermal water near Gila Bend. Fish are raised in geothermal water near Hyder.

A farmer in Marana irrigates citrus in winter with geothermal water. Hot springs near Tonopah, Safford, Willcox, and Mesa have been developed as spas.

Dairy farmers, fish farmers, and developers could be exploiting the same resource in Pinal County, he said.

"Basically, all around here, you've got geothermal," Pasqualetti said.

Source: www.kvoa.com/Global/story.asp?S=2823750&nav=J7NtVJsq, accessed May 31, 2005).

ALTERNATIVE FUELS

Alternative fuels are becoming a prominent part of the twenty-first-century energy paradigm. Shortly after the terrorist attacks on September 11, 2001, President George W. Bush announced a major policy initiative to promote the development of a so-called Hydrogen Economy—the use of hydrogen[7] for the production of energy, particularly for use in the transportation sector, in the United States, thus reducing dependence on the importation of petroleum (Moritsugu 2002). Hydrogen is used in fuel cell technology, which uses an electrochemical process of generating electricity.[8] Hydrogen can also be blended with natural gas to create a substance known as Hythane™, which reduces the emissions of natural-gas-powered engines and potentially extends natural gas reservoirs (Langreth and Fritz 1994; Sierens and Rosseel 2000).[9]

Ethanol, derived from corn, has long been a part of U.S. agriculture and energy policy and is considered a growth area for energy supplies of the future (Schulz et al. 2005). Ethanol and methanol[10] are blended with gasoline to produce lower emission combustion. Biomass—organic matter composed of various postharvest plant matter, animal waste, wood product waste, municipal waste, and the like—is a source for the production of biomass derived gases such as the aforementioned methane. Biomass is now used to produce approximately 3 percent of the U.S.'s "primary energy production" (www.eere.energy.gov/RE/biomass.html, accessed June 5, 2005).

Biodiesel is another alternative energy, derived from virgin seed oils (e.g., canola oil) or recycled cooking grease and other animal and vegetable fats. The diesel is a low-emission alternative to petroleum-based diesel and can be used in standard diesel engines without reducing power capacity. According to some sources, biodiesel can be produced by the layperson at their own home using plant and animal waste as a source (see www.biodieselwarehouse.com, accessed June 5, 2005; Pahl 2005). The DOE has sponsored projects to support technical and economic feasibility studies for biodiesel development. Projects have ranged from a study done at Middlebury College, Vermont to studies of soybean-based production in Tennessee. Production levels in Europe are already quite substantial, estimated at nearly two thousand metric tons in 2004 (www.ebb-eu.org/stats.php, accessed June 5, 2005).

PUBLIC PERCEPTIONS OF
ALTERNATIVE FUELS AND ENERGY[11]

A summary analysis of Roper Organization energy opinion polls is enlightening. In studies conducted between 1979 and 1989, Roper polls found that public support for alternative energy has held fairly steady, while there has been some decline in public support for the use of domestic coal as a replacement for foreign oil consumption. Solar energy was most likely to be identified as a method of achieving energy independence, followed by nuclear power and offshore oil. Hydroelectricity and wind power ranked slightly lower, with 20–30 percent identifying the energy sources as replacements for crude petroleum and natural gas imports. The public was aware of the increased costs associated with alternative energy, but the majority was willing to pay increased energy premiums for the opportunity to gain energy independence. Consistent support was offered for increased funding of renewable energy research and development, supporting a demand side solution to the domestic energy conundrum (see www.physics.pomona.edu/COURSESPhys17/papers/PubOp.pdf, accessed June 5, 2005).

In 2001, the Gallup Organization was uniquely positioned to conduct pre–September 11 and post–September 11 polls of U.S. citizen views of alternative energy. During the period of study May 2001 to November 2001, gasoline prices were approaching $2.00/gallon on a nationwide average. Since that time, the U.S. consumer has learned that gasoline and diesel fuel can become even more expensive. In the November study, Gallup was able to assess the impact of gasoline prices with the added fear of terrorism, a combined effect that led to increased support for energy independence.

The November 2001 Gallup Poll results found that there was a slight increase in support for opening the Arctic National Wildlife Refuge (ANWR) to oil exploration. In May 2001, approximately 38 percent were in favor of opening ANWR to exploration, but in the post–September 11 survey, 44 percent of respondents were in favor and 51 percent opposed the policy option. Support for nuclear power actually declined over the same time period: there was 48 percent support in May 2001 and 42 percent in November 2001. This change is attributed to a sense that nuclear power plants would be possible targets of further terrorist attacks on the United States. Thus, energy was being evaluated by the public not simply in terms of its cost, but also in terms of its relative "greenness" or availability, and in terms of it potential as a strategic vulnerability or strength. Wind, solar, fuel cells, and other forms of "new" energy were widely supported, with 91 percent of the survey respondents indicating that they were in favor of these sources of energy. Other options for America's energy future were also supported: the construction of additional power-generating plants (81 percent) and more fuel-efficient cars (77 percent). Alternative energy is a widely supported option, which, unlike nuclear power, is seen as both "green" and safe. Another conclusion that could be drawn from the 91 percent level of support found in the 2001 study is that the public sees these sources of energy as viable energy options. Despite the previous conclusion regarding post–September 11 concerns about energy safety, it is apparent in more recent studies that the public remains focused on alternative energy as important for the long-term protection of global climate health.[12]

Global public opinion is well captured in a comprehensive analysis of research compiled by the European Wind Energy Association (EWEA). EWEA is a trade association for the wind power industry, but the surveys compiled were conducted by independent public opinion polling organizations. The global surveys conducted in the European Union and Australia in 2002 and 2003 demonstrate a high level of respondent knowledge and significantly strong support for various renewable forms of energy production. The evidence is telling on another level because many Europeans and Australians have significant experience with the implementation of alternative energy policies, demonstrating significant public support as a result of successful energy policy innovations and positive implementation experiences. Practical knowledge seems to offer greater economic and political credence and policy legitimacy to alternative energy. Public opinion reflects the notion that alternative energy is something that is more than a distant concept: it has become and will likely remain an affordable and beneficial reality.

Additional time and experience with new generation energy systems may be required before the public accepts them as being safe and feasible. The use of hydrogen as an alternative energy/fuel, for example, has sparked some public comment because of its perception as being unsafe. When asked about the merits and detractions of hydrogen as a fuel, the word "Hindenburg" is often mentioned. A 2003 National Renewable Energy Laboratory study confirms that the public has very limited knowledge of relative safety of hydrogen compared to gasoline (Steiner 2003). A significant percentage of citizens have little or no knowledge of how hydrogen is used as a fuel. Nevertheless, the historic dirigible explosion of the Hindeburg in Lakehurst, New Jersey in 1937 remains to this day a focal point of American opinion regarding hydrogen.[13]

Public opinion studies lead to the conclusion that while public views about alternative energy are fairly positive, there needs to be further public education efforts to inform the citizen policy stakeholders about the direct and indirect benefits and costs associated with an energy source. Citizens must be encouraged to understand the process by which energy is created and how fuel is produced, stored, and dispensed. Safety-related issues should be also be discussed in greater detail and made understandable to nontechnical citizens. Public opinion that is soft in its level of support will be diminished by catastrophic events. Wavering political support for alternative energy policy would not serve the paradigm well when the unthinkable occurs and another energy disaster takes place. As noted previously, understanding is likely to also be a function of practical experience with new energy sources and energy storage methodologies.

CHAPTER SUMMARY

Alternative energy is defined at the federal level, but states often conceptualize it differently. It is important to have some unified sense of what alternative energy entails simply because public policy demands that target populations be monitored for successful and positive participation within policy areas, either through direct involvement or through more passive methods such as voting for a president.

The discussion of alternative energy sources, their production, use, and relative safety might lead one to conclude that there is no single energy policy. An understanding of what alternative energy sources exist as well as understanding the potential and weakness of all probable energy models allows for informed debate and choice regarding the adoption and use a viable energy paradigm. Informed public opinion in relation to alternative energy and sustainable communities will likely lead to greater consistency in policy implementation and outcomes.

NOTES

1. Note: In some cases, the exact bill specific definitions are helpful: "5) Low-impact hydropower, consisting of any technology that produces less than 50 megawatts of electric power and that harnesses the hydroelectric potential of moving water impoundments, provided such incremental hydroelectric development: (i) does not adversely change existing impacts to aquatic systems; (ii) meets the certification standards established by the Low Impact Hydropower Institute and American Rivers, Inc., or their successors; (iii) provides an adequate water flow for protection of aquatic life and for safe and effective fish passage; (iv) protects against erosion; and (v) protects cultural and historic resources." (Pennsylvania Senate Bill 1030 Section 2, Paragraph 5).

2. "(6) Geothermal energy, which shall mean electricity produced by extracting hot water or steam from geothermal reserves in the earth's crust and supplied to steam turbines that drive generators to produce electricity." (Pennsylvania Senate Bill 1030 Section 2, Paragraph 6).

3. "(7) Biomass energy, which shall mean the generation of electricity utilizing the following: (i) organic material from a plant that is grown for the purpose of being used to produce electricity or is protected by the Federal Conservation Reserve Program (CRP) and provided further that crop production on CRP lands does not prevent achievement of the water quality protection, soil erosion prevention or wildlife enhancement purposes for which the land was primarily set aside; or (ii) any solid nonhazardous, cellulosic waste material that is segregated from other waste materials, by-products of the pulping process and wood manufacturing process including bark, wood chips, sawdust and lignin in spent pulping liquors, SUCH AS waste pallets, crates and landscape or right-of-way tree trimmings or agricultural sources, including orchard tree crops, vineyards, grain, legumes, sugar and other crop by-products or residues. (8) Biologically derived methane gas, which shall include methane from the anaerobic digestion of organic materials from yard waste, such as grass clippings and leaves, food waste, animal waste and sewage sludge. The term also includes landfill methane gas. (Pennsylvania Senate Bill 1030 Section 2, Paragraphs 7–8).

4. "(9) Fuel cells, which shall mean any electrochemical device that converts chemical energy in a hydrogen-rich fuel directly into electricity, heat and water without combustion." (Pennsylvania Senate Bill 1030, Section 2, Paragraph 9).

5. One would hesitate to refer to this as a "credible commitment" on the part of government, attempting to establish the credibility of alternative energy. The phrase "credible commitment" was coined by Economics Nobel Laureate C. Douglass North

in reference to the commitment of Western European governments to the existence of property rights, something that may have played a significant role in the economic development of Western nations. As Timothy Frye succinctly states in his recent article in *American Political Science Review* (2004: 455):

> Central to this argument is the theoretical problem of credible commitment. Property rights are often weak because economic activities involve time-inconsistent exchanges between state and private agents. More precisely, laws and policies often promise benefits in the future for changes in behavior today. For example, to encourage investment, a government may pass a law promising tax benefits for 5 years for firms that invest. After a firm invests, however, it is vulnerable to ex post violations of its property rights my state agents. A government may impose confiscatory taxes regardless of legal rules to the contrary. Anticipating this possibility, rightholders will view their property as vulnerable and be reluctant to invest in the first place. . . . The problem of credible commitment underpins the irony that state agents with few constraints on their behavior may attenuate property rights precisely because rightholders understand that their property rights depend on the discretion of state agents.

A good example of this dilemma relates to the weak property rights of fossil fuel producers as they face increased government regulation and likely decreased tax incentives for fuel exploration and development in the future. In this book, therefore, I eschew the property rights debate, accepting the fact that property rights are—such as they are—weak or strong, depending on perspective. I am particularly interested in exploring what is the current state of the alternative energy policy arena.

6. Brine is salt water containing sodium chloride ($NaCl$), potassium chloride (KCl), calcium chloride ($CaCl_2$), and so forth. Some geothermal areas such as the geysers in northern California have very low salt content, but the vast majority of geothermal aquifers have high salt content.

7. Hydrogen is not commonly found as a free element. It has to be separated from other elements, often through the use of electrolysis or natural gas "stripping." Energy must be used to isolate hydrogen for use.

8. Incidentally, electricity, when used as a substitute for petroleum fuel in the operation of vehicles, is considered to be a form of alternative energy by the DOE (www.eere.energy.gov/EE/ trans_alt_fuels.html, accessed June 6, 2005).

9. Carbon dioxide is one form of emission that can be reduced through the use of Hythane™.

10. In the late 1980s, methanol was used in the development of a "problematic" yet low emission fuel known as M85 (see Bowlin 1999: 412).

11. Alternative energy opinion polls are commonly reported by interest groups and other advocates of alternative energy. As these polls might be viewed as biased in some manner, I chose to focus on studies conducted by nationally recognized unbiased polling organizations.

12. Thomas Brewer (2003) found that public support for alternative energy is related to public concerns about global carbon and other greenhouse gas emissions and that correlates suggest that public has consciously chosen policy positions and offered political candidates support on the basis of their relative energy and environmental

policy preferences (Brewer 2003 [http//www.ceps.befiles/ClimateDialogue/Brewer June03.pdf, accessed June 5, 2005]).

13. The Hindenburg explosion was not initiated by the hydrogen gas it contained.

WORKS CITED

——. 2005. *Public Opinion and Perception of Energy Technology and Policy*. www .physics.pomona.edu/COURSESPhys17/ papers/PubOp.pdf, accessed June 5, 2005.

BBC News. 2005. *Wind Farms Must Not Take Root in UK*. news.bbc.co.uk/1/hi/sci/ tech/4560139.stm, accessed November 26, 2005.

Bettelheim, A. 2000. Utility Deregulation: The Issues. *CQ Researcher* 10(1): 2–15.

Bowlin, M. 1999. Clean Energy: Preparing Today for Tomorrow's Challenges. *Vital Speeches of the Day* 63(13): 410–13.

Brewer, T. 2003. *U.S. Public Opinion on Climate Change Issues: Evidence for 1989–2002*. www.ceps.befiles/ClimateDialogue/BrewerJune03.pdf, accessed June 5, 2005.

Canadian Energy Association—National Resources Council 2002. "CEA-NRCan Alternative Energy Workshop: Summary and Possible Follow-Up Actions." www .canelect.ca/english/Pdfs/Summary%20of%20Proceedings%20from%202002%20 workshop.pdf, accessed May 26, 2005.

Energy Efficiency and Renewable Energy (EERE). 2005a. Glossary of terms, www .eere.energy.gov/financing/glossary.html, accessed May 26, 2005.

Energy Efficiency and Renewable Energy (EERE). 2005b. U.S. geothermal resource map, www1.eere.energy.gov/geothermal/geomap.html.

Frye, T. 2004. Credible Commitment and Property Rights: Evidence from Russia. *American Political Science Review* 98: 453–66.

Langreth, R. and Fritz, S. 1994. Hydrogen + natural gas=*hythane*. *Popular Science*. 244(3): 34.

Mandelbaum, David, and Brown, C. 2004. "Pennsylvania Energy Alert," Baltimore, MD: Ballard, Spahr, Andrews & Ingersoll, LLP. See www.ballardspahr.com/files/ tbl_s11Newsletters/PDFFile142/445/EnergyAlert11-04Final.pdf#search=%22 pennsylvania%20energy%20alert%20mandelbaum%22.

Moritsugu, K. 2002. Hydrogen Fuel Cell Technology Remains Many Years in Future. *Pittsburg Post-Gazette*, January 13, A-10.

National Wind Speed Technology Center. 2005. *Low Wind Speed Turbines*. www.nrel .gov/wind/about_lowspeed.html, accessed November 26, 2005.

Pahl, G. 2005. Heat Your Home with Biodiesel. *Mother Earth News*. Summer Special Issue: 57–63.

Physics Forum. 2005. *Power versus Energy*. www.physicsforums.com/archive/t-66458_ power vs. energy.html, accessed May 26, 2005.

Schulz, W., Wang, L., and Hess, G. 2005. Energy Bill Set. *Chemical & Engineering News* 83(31): 12.

Sierens, R. and Rosseel, E. 2000. Variable Composition Hydrogen/Natural Gas Mixtures for Increased Engine Efficiency and Decreased Emissions. *Journal of Engineering for Gas Turbines and Power* 122(1): 135–40.

Smallwood, K. and Thelander, C. 2004. *Developing Methods to Reduce Bird Mortality in the Altamont Pass Wind Resource Area*. Sacramento, CA: California Energy Commission.

Steiner, E. 2003. *Consumer Views on Transportation and Energy*. Golden, CO: National Renewable Energy Laboratory.

Tucker, Adrianne. 2005. *Middlebury College Receives $22,500 Grant for Wind Energy System*. web.middlebury.edu/offices/pubaff/news_releases/news_2005/20050511wind.htm, accessed May 31, 2005.

Toolbase Archives. 2005. *Solar Townhouses*. www.wholesalesolar.com/ products.folder/mount-folder/mount-info.html, accessed May 26, 2005.

Vestas. 2005. *Product Overview*. www.vestas.com/uk/Products/products2004/prodOverview_UK.htm, accessed November 26, 2005.

Wholesale Solar Electric. 2005. *Before You Choose: What You Need to Know about Solar Panel Racks and Trackers*. www.wholesalesolar.com/products.folder/mount-folder/mount-info.html, accessed November 26, 2005.

WEB SITES

telosnet.com/wind/early.html, accessed May 31, 2005.

www.awea.org/pubs/factsheets/avianfs.pdf, accessed June 1, 2005.

www.bbc.co.uk/norfolk/talk/wind_farms.shtml, accessed June 1, 2005.

www.biodieselwarehouse.com, accessed June 5, 2005.

www.ebb-eu.org/stats.php, accessed June 5, 2005.

www.eere.energy.gov/EE/ trans_alt_fuels.html, accessed June 6, 2005.

www.eere.energy.gov/RE/biomass.html, accessed June 5, 2005.

www.eia.doe.gov/cneaf/solar.renewables/page/solarthermal/solarthermal.html, accessed May 31, 2005.

www.kvoa.com/Global/story.asp?S=2823750&nav=J7NtVJsq, accessed May 31, 2005.

www.physics.pomona.edu/COURSESPhys17/papers/PubOp.pdf, accessed June 5, 2005.

www.psnh.com/Residential/ReduceBill/Applianceusage.asp, accessed May 31, 2005.

www.toolbase.org/tertiaryT.asp?DocumentID=1479&CategoryID=1093, accessed May 31, 2005.

www.ucsusa.org/clean_energy /health_and_environment/page.cfm?pageID=942, accessed June 2, 2005.

www.usda.gov/rus/regs/info/100-1/title_i.htm, accessed May 31, 2005.

www.vestas.com/uk/Products/products2004/prodOverview_UK.htm, accessed November 26, 2005.

Chapter Four

Alternative Energy/Fuels as a Public Policy Innovation

INTRODUCTION

Development and expansion of the alternative energy sector has been a function of both private sector and government-supported initiative, a historically typical scenario in terms of basic goods provisions and publicly accessed capital projects. The modern alternative energy paradigm and emergent public policy are milestones in the history of public-private basic goods and infrastructure development in the contemporary world of postindustrial and developing nations and markets. Energy is an essential global good and constrained supply would severely limit the public and private aspirations of individuals and the societies in which they live and work. This chapter will focus on alternative energy and fuel developments from the early 1970s to the present day.

POLICY INNOVATION AND ALTERNATIVE ENERGY

As noted previously, Presidents Nixon and Carter relied on top-down approaches to initiate policy redirection away from fossil energy and toward alternative and renewable energy development. Conversely, President Reagan approached energy policy from a free-market perspective, relying on marketplace solutions to energy needs and innovation. During the George H. W. Bush administration, energy policy innovation was further institutionalized through the passage of the reauthorized and amended Energy Policy Act (EPAct) in 1992. Building on EPAct, President Clinton adopted a middle-ground approach to alternative energy policy, combining the power of the national government and the concentration of technical expertise in national laboratories with the bottom-up policy innovations of state and local governments and communities to produce workable solutions for alternative energy production, storage, and use. Clinton's model of sound policy development is consistent with his commitment to a "laboratories of democracy" (Osborne 1990) approach to governance, engaging citizens at the grassroots level in the policy process.

Some Issues to Consider when Studying Policy Innovation[1]

Environmental Factors
  Is innovation driven by a real or perceived crisis?
  Are policy stakeholders satisfied with the status quo?
  Is the political culture of a jurisdiction conducive to policy innovation?
  Are the socioeconomic conditions in a jurisdiction favorable to innovation?
  Are adjacent states and local governments engaged in policy innovation? If so,
    what are the social and economic impacts of the policy innovations?
Ideology and Values
  Are elected representatives supportive of policy innovations?
  Are citizens supportive of policy innovation?

In studying the box above, effective innovation at the grassroots or state level requires circumspection for a reasonable chance at policy success. It is important to understand the context in which policy innovation occurs, the level of public support for innovation, and the level of citizen and institutional commitment to innovation success.

CONTRASTING STATE- AND NATIONAL-LEVEL POLICY DEVELOPMENTS

Policy innovations can be divided into six categories:[2]

1. Oil Shock I/Pre-PURPA 1970s (1973–1978)
2. Secondary Oil Shock Period (1979–1982)
3. Resurgence of Cheap(er) Petroleum and Growth of Deregulation (1983–1999)
4. Bush I and EPAct 1992 Reauthorization (1992)
5. Clinton and Post-EPAct 1992 Reauthorization (1992–2005)
6. New Millennial Demand Shifts and EPAct 2005 Reauthorization (2001–present)

OIL SHOCK I/PRE-PURPA 1970s (1973–1978)

The major state and local energy policy innovation in the century leading up to the oil shock were the public utilities commissions (PUC). The PUC model was initially established in many states in the mid- to late 1880s. Initially, the PUC regulated railroads in state and local areas and sought to stabilize suppliers and prices for railroad transportation. The PUC sanctioned a limited number of transportation providers and established prices for transportation, seeking to build stability in resource provision and availability.

With the advent of electricity, running water, and sewage in an increasing number of cities and towns, PUCs extended their power to sanction energy, water, and waste disposal service providers and to set prices for goods and services provided. Prices were arguably higher than if market forces had established prices, but the logic behind the PUC price setting was that through an essentially subsidy-based pricing of energy and other utilities, local and state economic growth was more likely to occur. The PUCs tended to be very closely aligned with utility providers, who are frequently motivated to increase profit margins. Price setting and the influence of the utilities in the PUC decision-making process led to decreased flexibility in the policy innovation process for energy as well other policy concerns—a circumstance that would prove damaging in the years following the 1973 oil shock because the model assumed regular growth in energy supply.

The oil shock placed state and local policy innovators, interest groups, energy industries, and citizen stakeholders in a quandary. Beyond a certain point, demand for energy is inelastic (see Olatubi and Zhang 2003). One needs to drive to work and to operate electrical utilities (e.g., heater, air conditioner, washing machine) at home. Demand-side policy innovations tried to educate citizens on methods of energy conservation. Certainly, informing citizens about turning lights off when leaving a room or keeping thermostats set at a higher temperature in summer are not politically charged innovations. Reducing demand for energy may not have had a significant effect on energy prices because supply was tightening simultaneously. Supply-side policy innovations were a bit more contentious.

The obvious point that emerges, of course, is that the post-1973 period faced a serious shortfall in supply of oil. But one should not confuse a shortfall in oil supply with a shortfall in potential energy supply. Oil shocks in 1973 did not present a serious supply crisis for energy in a general sense. The oil market of 1973 represented the fuels market reaching a new equilibrium point where consumers were required to think about their economic choices. In terms of public policy, 1973 represented the beginning of a debate about the nature of energy as a good—was it a private good or a marketable public good? If it were the latter, then supply of the good should be regulated and the nature of the good was best dealt with in a public forum. The policy debate in this period was about what could be categorized as politically and socially *acceptable* as an energy resource, as well as cost-efficient replacement fuels for transportation and electrical energy generation due to a tightening of petroleum supply, as well as a debate over citizen-consumer access to a resource.

A real or perceived crisis usually inspires some form of search process for potential solutions. Post-1973, states and local governments began the process of identifying viable energy resources available to make up for petroleum supply shortfalls. One of the earliest efforts was made in the state of California, which had started to establish stronger state energy policies in 1972 with the passage of the Miller–Warren Energy Lifeline Act (Southern California Edison 2005 [www.sce.com/CustomerService/understanding Baseline/history.htm, accessed June 11, 2005]). The act directed the California Public Utilities Commission to establish rate plans that would maximize energy availability through a graduated rate structure which charged higher prices to large consumers, while subsidizing the use of more limited energy users (an "'inverted' rate structure").

California energy policy encouraged conservation of energy so as to reduce demand on constrained energy supplies. The Warren–Alquist State Energy Resources Conservation and Development Act created a five-member commission that focused on both supply- and demand-based issues related to California's energy needs (California Energy Commission 2005 [www.energy.ca.gov/commission/overview.html, accessed June 11, 2005]). On the supply side, California possessed a multitude of state resources that could be used to meet energy demand. Petroleum resources in the state produced well over 60 percent of the state demand for crude petroleum. Nevertheless, oil prices are a function of global energy markets, constraining, among other things, the PUC energy model. When petroleum prices increase rapidly, its use in the generation of electricity would probably be cost ineffective because PUCs constrain electrical energy prices. Generating firms might have chosen not to produce due to mandated price ceilings and rising marginal costs but were contracted through PUCs to provide electricity regardless.

One method of reducing costs for electrical energy generation is to use cheaper energy inputs in the generation process. California governor Ronald Reagan looked to nuclear power as one source of cheaper inputs. Nuclear power was and remains a fairly controversial solution to electric energy production and supply. Environmental interest groups in particular launched an aggressive opposition to the nuclear option for meeting supply shortfalls and to reduce costs to energy producers and consumers, reinforcing the point made earlier that the success of innovation is a function of citizen acceptance. Opponents of nuclear energy proposed alternatives, seeking to gain public support and establish a new energy paradigm built on environmental concerns, concluding that the nuclear power option was environmentally, politically, and socially untenable (see Cochran et al. 1975; *New York Times* 1970: 41). Proposed energy alternatives often focused on the development of "clean" renewable energy alternatives, such as wind and solar technologies. Solar and wind energy solutions faced significant challenges. First, the technologies were, for large-scale commercial purposes, still underdeveloped. Individual consumer use of these clean energy sources was possible, but would likely not produce the energy supply needed to meet consumer demand unless demand ws reduced. The second major problem was related to the first: namely, the cost of clean energy sources to be used for commercial purposes was—at that time—prohibitively expensive (see Cupulos 1979: 159).

In 1977, the California Innovation Group (CIG), a local-level policymaking body operating on financial assistance from the National Science Foundation, developed a collaborative relationship with Energy Research and Development Administration, the latter authorized by Congress in 1974 under 42 USC 73 Ch. 1 §5811 and intended to aid local and state governments responding to the 1973 energy "crisis." The CIG, composed of eight California municipalities, researched energy supply options related to geothermal/coal hybrid generation projects and developed plans to reduce peak hour energy consumption. In their article detailing the CIG, Michael Conway and Gregory Simay (1977) provide a cogent case study of policy innovation occurring in California and in other states in the early 1970s: namely, an active search for alternative sources of energy and the broadening of a discussion about what sources of energy would be most acceptable to consumers in terms of cost, availability, and relative safety:

> Many approaches to save [and produce] energy have merit. However, to enjoy public support, they must be carried out in harmony with the total social and economic environment of the nation's cities. In particular, energy conservation and development programs must be successful when applied to cities of medium size [author note: the vast majority of cities in the United States]. (Conway and Simay 1977: 711)

The emergent model during this period emphasized increased state and local dialogue about the specific energy demands and supply issues related to particular state or local area needs. The California case illustrates well the 1970s energy policy period, demonstrating the growing emphasis on devolving planning and responsibility to the state and local level. In terms of intergovernmentalism, the federal role at the time was to help states become more aware of their relative energy self-sufficiency or vulnerability and to encourage the development of long-term planning to meet the current and future energy needs through a more diversified energy mix (see Sawyer 1984) along with the continued and historical federal regulation of energy markets (see www.ferc.gov/, accessed June 11, 2005).

National energy policies in the post-1973 period continued to focus significant attention on intergovernmental solutions, although employing primarily top-down methodologies. The 1973 oil shock was initially considered by a price control implementer. President Richard Nixon had spent time during World War II working in a price control policy office operated by the U.S. Navy. Building off his wartime experience, Nixon had responded to inflationary trends that had arisen in the latter part of his first presidential term with a price control policy innovation; his response to the oil shock differed only slightly. The Ford administration maintained the price control strategy as emphasized in the Energy Production and Conservation Act of 1975, occurring in the same year as the first major institutional change in national energy policy—the establishment of the Energy Research and Development Administration (*Economist* 2005a [www.25yearsofenergy.gov/origins.html, accessed June 13, 2005]). President Carter continued the price control strategies of Nixon and Ford as methods of alleviating some of the cost strain on consumers, but took a slightly different tack through the development of a cabinet-level response—namely, the creation of the U.S. Department of Energy (DOE). President Jimmy Carter expanded energy policy into an intergovernmental response detailed above and through the use of grants and national research and development efforts to increase energy supply long-term through alternative sources of energy generation and storage. Carter also sought to reduce energy demand through legislation mandating more gas-efficient automobiles.

Large-scale coordinated intergovernmental innovation efforts—and success stories—were limited for several reasons. First, the post-1973 period was characterized by problem definition and solution identification. Second, national political priorities lacked clarity due to political turmoil at the presidential level and the installation of an unelected presidential administration. Third, oil prices, as a measure of "crisis," increased only marginally until midway through the Carter presidency.

International cooperative efforts also developed during the post-1973 shock. In a study conducted in late 1970s, Bobrow and Kudrle (1979) found that international coordination of energy policy research and development (R & D) was an important part of global energy response to supply shortfalls. The biggest challenge to coordinated effort, however, was international rivalry for resources and information. Perhaps equally important, the study found that comparative analysis of R & D budgets in the Western democracies of Europe and North America yielded significant per capita variation in expenditures. In other words, commitment to developing alternative and other future energy resources was not equally shared and resource development strategies, at times, quite divergent, which meant that cooperative effort was strained (see Bobrow and Kudrle 1979: 156).

SECONDARY OIL SHOCK PERIOD (1979–1982)

The oil shock of 1973 leveled off in terms of impact on fuel prices, giving consumers some sense that the crisis had, at the very least, entered into a period of status quo. Prices were still high, but at least they were not getting much higher. Shortly after becoming president and establishing the Department of Energy, the Carter administration was faced with a secondary oil shock between 1978 and 1980. The price of oil shot up nearly 90 percent, to an annual average of $21.65 per barrel in nominal dollar terms (approximately, $49.63 in 2005 dollars). The status quo was unacceptable to policymakers and other policy stakeholders, which was reflected in President Carter's low public approval ratings at that time, albeit there were several confounding factors that shaped his approval ratings (e.g., the Iran hostage crisis).

The political ideology of the time was influenced by the general value structure of policymakers. Members of Congress and the bureaucracy had experienced the regulatory and distributive policies of the Franklin D. Roosevelt (FDR) administration, which had succeeded in alleviating the impact of the Great Depression and bringing the United States successfully through a world war. A new generation of policymakers—predominantly baby boomers, were introducing the green energy paradigm. The combined effect of generational experiences, ideology, and sociopolitical value frameworks produced a new energy policy solution known as PURPA—the Public Utility Regulatory Policies—Act of 1978.

PURPA was a major national policy commitment to diversifying the nation's energy sources. The act established national standards for energy policy—an aspect of the new policy that was particularly top down as an innovation. While utility regulation had historically been controlled by state commissions, whose meetings were attended almost solely by power generation and transmission engineers and executives as well as commission members, PURPA opened state energy policy meetings to greater public participation and input on energy policy. PURPA also sought to increase the level of equity in terms of fair energy provision and pricing.

Under the federal law, state utility commissions were encouraged to use third-party power generators so as to increase supply and lower costs. Many of these third-party power generators had excess supply available for use at various peak or nonpeak times; using these power sources would increase energy efficiency. Several third-party power generators used "alternative" and/or renewable energy sources in generating electrical energy, such as natural gas, solar, and wind power. For many of its proponents, the inclusion of public input and third party generation made PURPA the embodiment of cooperative federalism—a policy that recognized the value of bottom-up approaches to achieving policy solutions (see Salisbury 1980). It was and continues to be seen as an important attempt to recognize the importance of unified energy policy in order to maintain a national economy in times of energy uncertainty. Additionally, many aspects of PURPA reintroduce the importance of social and economic equity in the provision of electrical energy in the United States. Although many aspects of PURPA were regulatory, it was viewed as only one aspect of a new energy policy paradigm that emerged in the 1970s. Many other aspects of energy policy, such as the development of national energy laboratories (to be discussed later in the chapter), were seen by proponents of PURPA as distributive policies designed to help states and local governments cope with the retooling required as fossil fuel-based economies began to shift toward alternative energy.

Energy safety issues also characterized this policy period, effectively limiting the viability of nuclear energy as a future solution. The Three Mile Island incident on March 28, 1979 (see Janson 1979), in which nuclear radiation was vented into the air from an overheated reactor core, made nuclear power a politically unpopular alternative energy source. While nuclear energy continued to play a sizeable role in the U.S. energy supply, the construction of new power plants became politically unfeasible due to the image—reinforced by media and by ecology-minded interests—of nuclear power as dangerous to individuals and to the environment.

RESURGENCE OF CHEAP(ER) PETROLEUM AND GROWTH OF DEREGULATION (1983–1999)

Reagan: Policy Environment and Innovations

The election of Ronald W. Reagan in 1980 occurred within one year of the "high tide" of oil prices. In 1981, crude petroleum prices reached an annual average of $66.20 per barrel (in 2005 dollars). Energy policy was quite naturally a major issue on the policy agenda. Elected on a variety of issue positions, Reagan had campaigned on the issue of energy policy and proposed solutions at odds with many of the renewable energy and regulatory policy innovations implemented in the previous decade.

Following Reagan's inauguration in January 1981, many aspects of the alternative energy paradigm began to give way to changing economics and political values. Reagan focused on both demand and supply-side issues related to fossil fuel energy policy:

- "Decontrol" (Executive Order 12287) fuel prices
- Increase access to domestic oil exploration in Alaska
- Increase access to natural gas exploration and transportation domestically
- Promote "alternative" energies, primarily nuclear energy

The logic of Reagan's model of innovation in energy policy was based on his belief in free-market economic principles as the optimum solution to the energy "crisis."

Lifting controls on petroleum prices means that the government would not longer place price limits on the price of petroleum sold to consumers, an innovation heavily promoted by former Republican president Richard Nixon (see Smith and Phelps 1978: 428). Certainly the academic literature and practical experience had shown that price controls have the potential to produce unintended effects on consumer behavior (see Shultz and Dam 1977: 151; Weber and Mitchell 1978). Price controls may send false signals to markets and consumers, resulting in the overconsumption of a good or service. From Reagan's perspective, clamping down on supply and controlling consumer prices and distribution was simply not the answer: either let price signals shape consumer choice or increase supply.

Laws such as PURPA, while perhaps viewed as positive in terms of opening up the energy market to third-party producers, were not seen as beneficial by the Reagan administration in terms of long-term energy solutions, particularly if energy costs associated with third-party producers stood at levels far above market equilibrium price at any given point in the future. The combined effect of PUCs and PURPA would force energy providers into high-cost energy choices while simultaneously limiting prices charged to consumers. Prices shaped by these energy-related policies were seen as false, causing consumers to overconsume, while negatively impacting utility companies. In the 1980s, and in the post-Reagan period of the 1990s when crude petroleum and other carbon-based energy sources stood at record low prices, the notion of paying for alternative/renewable energy supplies, such as solar or wind, was an economically undesirable effect of PURPA and its progeny. In a very general sense, PURPA's third-party provider concept was good in theory but misapplied in practice.

Third-party producers were seen by Reagan and others as being part of a long-term solution. Third-party producers would become a part of a more expansive energy market when prices made renewable energy economically feasible. In the meantime, however, there were cheaper and quite possibly simpler energy alternatives to the supply issue. Domestic crude petroleum exploration was one method of short-term increases in supply and the North Slope of Alaska was seen as one possible solution to the supply problem.[3] Reagan also continued to promote many of the "alternative" energy policies that he had been promoting since his governorship in California. Nuclear energy was viewed as a viable solution to future energy shortfalls. Given that per kilowatt costs for nuclear energy were comparable or even lower than other forms of energy generation and that potential liability costs had been lifted from producers' shoulders via the Price-Anderson Act (see Munson 1979), nuclear energy was an attractive option for energy supplies of the future.

BUSH I AND EPAcт REAUTHORIZATION (1992)

The politics of energy policy changed in the years following the Reagan presidency. Although his influence was arguably so strong that a return to 1970s policy innovations was highly unlikely, a new model emerged. The values of the new model are in part based in the PURPA model but were also shaped by a more pragmatic, so-called Third Way, policy paradigm. The major energy crisis that occurred during the presidency of George H. W. Bush was a circumstance related to the Persian Gulf War. While theories abound about the war being about oil, a more objective way of viewing the conflict is that it illustrated a connection, perhaps intuitive: Middle East political stability and motivations could negatively impact the world supply of petroleum. Although the annual average price of petroleum did not increase during the Persian Gulf War period, the war did cause some policy debate at the national level over the need to further diversify the nation's energy portfolio—an idea that state and local governments had promoted for years through bottom-up policy innovations.

A major energy policy innovation of the George H. W. Bush presidency, perhaps the least-remembered aspect of his term of office, was the reauthorization of the EPAct of 1992. Through the reauthorization process and amendments, national policymakers reaffirmed a commitment to energy source diversification. More strongly supported than in the original authorization, EPAct of 1992 created a national priority for the creation of clean *and* sustainable energy options for communities: in effect, it was a commitment to lower energy emissions, cleaner air and water, and sustainable energy sources for the purposes of "livability" in the broadest possible sense. The national priority was framed in a manner that moved the nation beyond the price-control versus free-market debate.

EPAct of 1992 (1993 P.L. 102–486) significantly altered rules and incentives for the energy market, setting the stage for energy policy innovation during the Clinton and George W. Bush administrations. From a rational choice perspective, it is important to understand several important aspects of EPAct. Many aspects of public law represent meaningful and long-lasting commitments to a policy direction and a set of core values central a policy "game," impacting the choices of the participant players. Below is a brief summary of the core sections of EPAct of 1992 and the Clean Cities paradigm, both of which tie energy to environmental quality issues:

- Energy Efficiency
  Amended the *Energy and Conservation Standards for New Buildings Act of 1976* (see 42 USC §6831)[4] to include an increased emphasis on voluntary energy efficiency standards created by "consensus" among various building and heating/cooling engineering organizations (e.g., Council of American Building Officials [CABO]). Federal, state, and commercial building energy efficiency codes are to be standards set in a manner that would either meet or exceed CABO guidelines, as reviewed by the DOE. Manufactured home energy efficiency standards were raised and the Department of Housing and Urban Development was tasked with reviewing those standards. Energy-efficient homes were encouraged through a mortgage guarantee pilot program that was tied to the National Housing Act (42 USC 12832). Amendments were made to the Public Utility Regulatory Policies Act of 1978 (16 USC 2601), encouraging energy utilities to develop comprehensive integrated energy planning; the amendments also fostered a role for small energy producers. The efficiency section also provided efficiency grants to state and local agencies and utilities involved in energy regulation and production. Finally, the efficiency section of the act encouraged the development of efficient appliances and equipment for commercial and residential use.
- Natural Gas
  Amended the Natural Gas Act of 1978 (15 USC 717) and emphasized the need to create a free market for natural gas and liquefied natural gas being imported into the nation.

- Alternative Fuels—General
Defined alternative fuel as:

> methanol, denatured ethanol, and other alcohols; mixtures containing 85 percent or more (or such other percentage, but not less than 70 percent, as determined by the secretary, by rule, to provide for requirements relating to cold start, safety, or vehicle functions) by volume of methanol, denatured ethanol, and other alcohols with gasoline or other fuels; natural gas; liquefied petroleum gas; hydrogen; coal-derived liquid fuels; fuels (other than alcohol) derived from biological materials; electricity (including electricity from solar energy); and any other fuel the secretary determines, by rule, is substantially not petroleum and would yield substantial energy security benefits and substantial environmental benefits.

The section also clarifies that alternative fuel vehicles are to be considered as such *if* they are manufactured as alternative fuel vehicles and that at least 50 percent of the alternative fuel vehicles are to come from domestic (United States and Canada) vehicle manufacturers. The federal vehicle fleet was directed to become at least 75 percent alternative fuel powered by 1999 and develops a plan for refueling sites and operations and maintenance (O & M) to be completed per manufacturer standards.

- Alternative Fuels—Nonfederal Programs
This section encouraged the use of alternative fuel vehicles for commercial and for state and local governments. The section directed the U.S. secretary of energy to develop and coordinate a public information campaign about the costs and benefits of alternative fuel vehicles. The section also encouraged the states and local governments to offer tax breaks to individuals and organizations that purchase alternative fuel vehicles. Section 410 encouraged the development of alternative fuel bus systems in cities and states coordinating the efforts of DOE and Department of Transportation (DOT). The DOT was also directed to provide financial assistance to states and local communities that were encouraged to retrofit school buses or purchase new alternative fuel buses with the intent of reducing air emissions—a $30 million budget appropriation annually was provided in EPAct 1992 for up to three years. Finally, a low-interest loan program was established for small businesses for the purchase of alternative fuel vehicles.

- Availability and Use of Replacement Fuels, Alternative Fuels, and Alternative Fueled Private Vehicles
This section of EPAct of 1992 effectively mandated the use of alternative fuel vehicles for any "person"[5] involved in the production, transportation or sale of energy. Additionally, any large consumer of petroleum energy (i.e.,

over 50K bbl/d) were required to use alternative energy vehicles. The section also detailed the need for the production of alternative energy, which in many cases involves agriculture (i.e., corn) and required the DOE to monitor and seek to balance the availability and use of alternative fuel. Federal, state, local, and private fleet vehicles for persons as defined above were required to expand their alternative fuel capacity to 70 percent of all fleet vehicles by 2006.

- Electric Motor Vehicles
  Defined electric vehicles as "a motor vehicle primarily powered by an electric motor drawing current from rechargeable storage batteries, fuel cells, photovoltaic (PV) arrays, or other sources of electric current and may include an electric-hybrid vehicle" ["a vehicle primarily powered by an electric motor that draws current from rechargeable storage batteries, fuel cells, or other source of electric current and also relies on a non-electric source of power"] (Title VI §601). EPAct of 1992 directed the DOE to request proposals for demonstration projects promising development of more effective and efficient electric vehicles. The section also encouraged collaboration of manufacturers in the development of an electric vehicle infrastructure, but sought to assuage concerns regarding the use and distribution of industries' proprietary information.
- Electricity EPAct of 1992 maintained requirements that electricity wholesale generators receive Federal Energy Regulatory Commission (FERC) recognition; but amended Public Utility Holding Company Act (PUHCA) of 1935 and increased state-level power in terms of meet PURPA alternative energy goals. EPAct of 1992 also amended the Federal Power Act and changed interstate transmission rules so to loosen electric wholesale generators' control of transmission lines and prevented the use of price setting to reduce access to third-party electric wholesale generators' transmission capability using existing transmission infrastructures.
- High-Level Radioactive Waste; U.S. Enrichment Corporation; Uranium Revitalization; Uranium Enrichment Health, Safety, and Environment Issues
  Within EPAct of 1992 legislation, Congress quietly loosened the regulatory process for the establishment of new nuclear power plants (*Congressional Quarterly* 1992). "Alternative" nuclear energy was discussed in EPAct of 1992, based on the concept of using lower-radiation reactors to generate power, while improving the safety of nuclear energy for communities. Amending the Atomic Energy Act of 1954 (42 USC 2011 et seq.), Congress created the United States Enrichment Corporation, which was tasked with, among other things, overseeing the recycling of spent weapons grade fissile materials for use in civilian nuclear reactors (see 102 Cong. H.R. 776 §901). While producing a significant portion of the nation's electrical

power, the industry was under severe strain because of significant limitations on its ability to grow and even to maintain its existing infrastructure.

- Renewable Energy
  This section of EPAct 1992 was designed to provide grant opportunities for
  further research, development, and demonstration of the use of renewable
  energy resources. The section effectively defined renewable energies as
  biomass, ethanol, biofuels, PVs, solar thermal, wind energy, geothermal,
  fuel cells, and nondefense superconductors.
- Coal
  Under §1301–1305 of the act, Congress directed that further research be
  conducted into the development of clean coal technology (i.e., reduced carbon emissions). Such technologies could be used to develop coal-based
  diesel for combustion engines and as fuels to operate turbines in electricity
  generation. In §1306, Congress calls for further research and development
  of natural gas recovery from coalbeds (also in §1309). Coalbed wastes
  should also be used an energy source.
- Global Climate Change
  Adopting the findings of a National Academy of Sciences report on global
  warming, Congress directed DOE to produce a "least cost energy strategy"
  that looks at the full range of benefits and costs associated with a variety of
  energy mixes (i.e., fossil fuel, alternative, and renewable energy).
- Reduction of Oil Vulnerability
  Under Title XX of EPAct of 1992, Congress stated that the energy supply
  and the consumption of energy is an integral part of an efficient economic
  system. Economic performance based on foreign energy supplies increases
  economic vulnerability and may cause national security weaknesses, too.
  Additionally, greater emphasis on clean fuels will benefit the environment
  and will likely improve public health.
- Energy and Environment/Energy and Economic Growth
  In many ways, Titles XXI and XXII provide further details on various alternative and renewable energy program initiatives discussed earlier in
  EPAct of 1992; a few other energy-related programs are discussed regarding strategic metals, for instance. Title XXII reaffirmed the role of alterna-
  tive/renewable energy technological development as an important future
  growth area for the domestic economy.

CLINTON AND POST–EPAct 1992
REAUTHORIZATION (1992–2005)

The Clinton years witnessed a period of relative tranquility in terms of crude
petroleum prices. A lack of an energy "crisis" could easily have reduced a

commitment to alternative energy development. What may have led to continued innovation in energy policy, however, was the particular set of core values upon which Clinton's administration was built. A young governor of the state of Arkansas at the time of PURPA's passage, Clinton had been exposed to the energy crisis and its impacts on states and local communities for most of his career. Reelected governor of Arkansas in 1982, he had served in that capacity during a decade of Reagan-Bush leadership in energy policy innovation. He remained committed to diversification of the energy portfolio, demonstrably during his presidency in the implementation and funding of EPAct of 1992.

Clinton's personal ideology focused on pragmatic solutions and thoughtful policy experimentation, utilizing to a significant degree the power of intergovernmentalism. Bill Clinton and his ecologically-minded vice president, Albert Gore Jr., worked with state, local, and federal agencies to cooperatively explore green energy solutions to balance environmental and human community needs. A significant difference from pre-Clinton policy innovations was the optimistic aspect of the policy innovations advanced. In the 1970s and 1980s, much of the energy policy focused on energy independence or energy sustainability. In essence, policy took on a defensive posture; as a policy innovation, the general milieu of the energy policy debate was seemingly driven by a need to protect postindustrial predominantly Western economies against an uncertain energy future. Clinton and Gore, however, saw an opportunity to build a new economy around the green power model. For evidence of policy innovation success, they pointed to European nations such as Germany's green power infrastructure and its positive impact on that nation's economy. Additionally, academic evidence tied environment quality to improved global economies; the concepts served to reinforce the value of the Clean Cities policy innovation. As with all policy innovations, however, optimism is often tempered with the need to focus on contingencies and changing realities (see *Economist* 1992: p. 18); but post-Clinton energy policy events have forced U.S. policymakers to focus on the need for accelerating the energy policy innovation process, whether the outcome will be predominantly market controlled or government managed.

The Clinton administration emphasized presidential executive order authority rather than through legislative action. In 2000, President Clinton signed E.O. 13149, entitled *Greening the Government through Federal Fleet and Transportation Efficiency*. The order required that federal agencies take a "leadership" role in promoting cleaner burning vehicles by reducing demand on petroleum. Clinton required agencies to reduce fleet fuel demand by 20 percent (in terms of 1999 consumption). Part of this could be achieved by simply reducing the distances traveled by vehicles in a fleet, but Clinton ordered agencies to pursue the purchase of Alternative Fuel Vehicles (AFVs) and more economical vehicles in terms of fuel efficiency, size, and durability in relation to use. By the end of 2005, as stated in the E.O., the majority of fuel demands by federal agency fleets shall be met by alternative fuels. Additionally, the new vehicles would need to abide by the EPA's Tier 2 standards with regard to vehicle emissions.

NEW MILLENNIAL DEMAND SHIFTS AND EPACT 2005 REAUTHORIZATION (2001–PRESENT)

Without question, the terrorist attacks of September 11, 2001, and military engagement in Iraq served as important motivating factors in shaping the energy policy of the twenty-first century. Prior to September 11, President George W. Bush advocated the opening of the Arctic National Wildlife Refuge (ANWR) to further oil and gas exploration. Recognizing mixed public opinion regarding petroleum exploration in the ANWR (see Lazar 2001: 1), Bush actively promoted alternative energy sources, such as hydrogen, in the period leading up to his reelection bid in 2004. The president's Hydrogen Initiative was an attempt to accelerate the movement toward a hydrogen-based infrastructure in the United States.[6]

The initiative built on the Clinton administration's hydrogen policy innovations in substantial ways. President Bush's plan was a multiyear research and development to advance knowledge and development of hydrogen fuel and fuel cells. Additionally, the initiative sought to develop a nationwide hydrogen-fueling infrastructure. Safety issues for the mass production, transportation, storage, and use of hydrogen were important factors studied by federal laboratories, university scientists, and private sector research and development teams. The figure below summarizes the systematic analysis of converting an entire society constructed around carbon-based energy production and use, into a society based on hydrogen as an energy source. Each one of the major topical areas in the figure below (figure 4.1) will require vast amounts of research and knowledge collection before knowledge can be transformed into a series of practical policy innovations. In many ways, the Hydrogen Initiative was a major first step in reconfiguring an entire society and economy in the new century. New codes and standards for safety and use will need to be constructed, but on the basis of sound science paired with a shared set of social and political values regarding energy policy.

In a report released in January 2003, the National Research Council (NRC) concluded that more research and development was needed before the ambitious goal of building a hydrogen based economy could realistically be attained—quite possibly, decades from the present day. Based on NRC find-

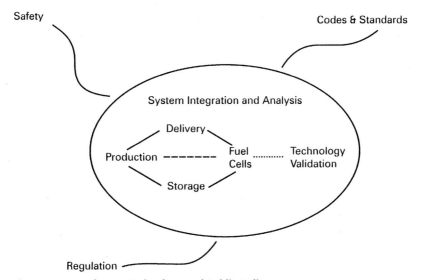

Figure 4.1. Hydrogen Technology and Public Policy
Source: Redrawn by Christopher A. Simon; an adaptation from www.eere.energy.gov/hydrogenandfu-elcells/mypp/pdfs/systems_analysis.pdf, accessed June 15, 2005.

ings, the Hydrogen Initiative is still a project in its infancy. In the meantime, however, the Energy Efficiency and Renewable Energy Program within DOE, partnering with state and local government as well as private sector researchers, will proceed with the goal of creating the energy paradigm of the future.

As oil prices continue to rise in the early twenty-first century, however, there is, as yet, no large-scale replacement energy paradigm in place. Energy demands continue to increase as the economy of the nation continues to grow at a vigorous pace, especially when compared with other industrialized nations. In an effort to bridge the gap and to increase energy supplies, President Bush has focused renewed attention on ANWR oil and gas exploration. In early 2005, Congress voted to open a small section of ANWR to further oil and gas exploration; perhaps a small step in an incremental process of opening a larger portion of ANWR in the future. As a long-term policy solution, further reliance on Alaskan oil and gas may not have the intended policy impact of slowing down short-term innovation by moving supply curves to the right, lowering per unit energy prices.

Regardless of petroleum supply, the oil and natural gas processing and production infrastructure are limited in capacity, which increased supply is a function of access. The impact of Hurricane Katrina along the Gulf Coast of the United States in 2005 demonstrated that supply issues and production capacity issues are inextricably intertwined. Additionally, the production paradigm is driven by market forces and government regulation.

Until quite recently, the resource prices kept many sources of petroleum from the market simply because they were not profitable. Good examples of formerly underutilized sources of petroleum are the oil sands found in Alberta, Canada, and heavy sour petroleum (sulfur rich and highly viscous) used by the United States and many countries as the basis of fuel oil. Heavy sour crude is readily available and because of demand has a current price less than one-half the price of sweet light crude. The current price of sweet light crude has made profit margins for the development and refinement of heavy sour crude more attractive to the oil industry and has made retooling refinery capacity more cost effective for the processing of heavy sour crude.

With the issues of supply and processing in aspects of the extant and future-oriented energy supply, Congress passed and President George W. Bush signed into law the Energy Policy Act of 2005. The 2005 reauthorization is comprised of series of energy policy innovations designed to meet a changing energy policy paradigm in terms of petroleum availability (i.e., innovations intended to restructure energy supply) and a series of regulations to force change in energy demand structures that would lead to further development and use of AFVs as well as other forms of alternative energy.

EPAct of 2005 has changed the energy paradigm in significant ways. A number of energy regulatory structural changes occurred as a result of the Act. PUHCA, for instance, was not reauthorized, its duties being transferred to Federal Energy Regulatory Commission (FERC) and to relevant state level administrators. PURPA was reformed to make it more cost effective for new energy providers to increase electrical energy supply. EPA clean air regulations are tightened by EPAct of 2005. The act makes no accommodation for a national-level Renewable Energy Portfolio, leaving that policy innovation to the voluntary implementation by various states.

Alternative energy is offered two major concessions. First, the DOE remains committed through the principles of its initial authorization as well as through goals established in EPAct 2005 to continue to conduct research and demonstration project development in the area of alternative energy. Second, and more specifically, EPAct 2005 delineates a series of business and residential user tax credits for the use of alternative energy systems. The tax credits have generally increased from 10 percent of initial project development costs to approximately 30 percent. Increased use of tax credits is indicative of a greater commitment to the development and use of alternative energy.

In another development, conceptualization of alternative/renewable energy is broadened through statutory application. First, next generation nuclear energy plants are offered incentives for development and operation. EPAct 2005 specifically calls for a demonstration project of a next generation nuclear power plant to be developed at Idaho National Laboratories within the first quarter of the current century. The law also contains stipulations for further study and development of nuclear waste processing and the further identification and development of waste depository sites. Second, hydropower, which was the subject of much criticism by ecologists and environmental groups for its purported impact on wildlife and plant species in the Pacific Northwest and elsewhere, was provided a major boost by statutory commitment to the further development of hydropower energy production efficiencies. The law also offers hydropower developers a significant role in defining the conditions under which development occurs as well as defining alternative impact studies with recommended alternative solutions to deal with externalities (e.g., hydropower dam impact of fisheries and plant life).

There was a strong commitment to alternative energy in the president's Hydrogen Initiative. Incentives are in place for the development of alternative fuel production sites. Production of fuels such as hydrogen will be accelerated by the development of next-generation nuclear power, resulting in cheaper hydrogen as compared to hydrogen produced using renewable energy sources such as wind and solar.

EPAct 2005 makes a very strong commitment to the further development of domestic petroleum and gas. Energy companies are allowed to reduce their royalty payments, which will result in lower costs and greater gross (and possibly net) profit margins. Additionally, EPAct 2005 directs the DOE and affected states to conduct a thorough review of petroleum reserves, particularly offshore reserves. It is likely that increased off shore petroleum fields will be developed as a result of these studies. The law also commits the nation to further development of natural gas and petroleum pipeline and refinery development. Clean coal development and equipment are offered tax incentives. Liquefied natural gas development is also advanced by EPAct 2005 and will be managed by FERC in cooperation with appropriate state level bureaucracies. Absent from EPAct 2005 is any further discussion of the opening of the Arctic National Wildlife Area to petroleum and natural gas exploration.

Energy efficiency is perhaps less of a focus of the bill than energy development. For instance, there are no new requirements for increased automobile efficiency. Yet, EPAct 2005 does set standards for the near doubling of ethanol supply to 7.5 billion gallons by 2012. As a source of oxygenation for gasoline, ethanol has been shown to significantly reduce CO emissions in the combustion process (Schifter et al. 2001).

Promoting Equity through Energy

One of the enduring themes in federal, state, and local energy policies in the twentieth and twenty-first centuries is the goal of achieving equitable energy solutions. The exact meaning of the word "equitable" has, of course, evolved over time. In the 1930s, FDR created the blueprint for future discussions. The Roosevelt administration's energy policies focused attention on price of energy as well as availability of energy. Energy prices were relatively high prior to the Roosevelt administration, largely because the costs of developing a privately owned and operated energy infrastructure, as well as the costs of producing energy, were exorbitant. Electric energy was available primarily in large urban areas, which used government policy influence and incentives to aid in the establishment of an energy infrastructure. Fuels, such as gasoline, were also expensive. Automobile costs had declined as a result of mass-assembly plants developed by Henry Ford, but fuel was not cheap, nor was it always easy to access.

Roosevelt's policies focused on electricity infrastructure and production. His vision was a society that would have cheap electricity and a solid infrastructure to transmit electricity. As will be discussed in subsequent chapters, hydropower was the primary method promoted by Roosevelt. With a solid electricity infrastructure, the economy would grow and automobiles and gasoline, among other seemingly "luxury goods" of the time, would become more readily available to consumers. But electrification was the primary policy innovation used to promote equity in society and in the marketplace.

The late twentieth century brought a renewed focus on equity, but one that really expanded the meaning of the term. As energy policy and environmental policy issues converged, the concept of equity expanded beyond accessibility and price-related issues and began to focus on the impact of energy policy on the environment and on public health. In terms of environmental issues, so-called greenhouse gases became a more prominent issue in the policy equity debate. Additionally, scientific evidence has increasingly shown the impact—at times quite uneven across society—of carbon-based fuel emissions on human beings. The evidence shows that children and individuals of lower socioeconomic status are more likely to be exposed to and harmed by carbon-based fuel emissions. Thus, the issue of energy policy equity remains focused on the original issues of availability and cost, but increasingly, there is a focus on the impact of energy types on citizens and the environment. In that sense, zero emission energy sources are more consistent with the broader definition of policy equity in relation to energy policy.

CHAPTER SUMMARY

A major shift in energy policy innovation began at the state level with the passage of the Miller-Warren Energy Lifeline Act of 1972. It was not until the 1973 oil shock that significant federal innovation began to take form—as policymakers became increasingly aware of the need to reconsider the national energy policy paradigm that had been in place since the PUHCA of 1935. Federal policymakers in the 1970s realized the need to include state and local policy innovators, as well as private stakeholders, in the energy policy innovations of the future. An increasing supply of oil in the late 1980s and for much of the 1990s may have reduced national attention on the need to enlarge and diversify the national energy portfolio, but state-level initiatives that emerged from PURPA of 1978, EPAct of 1992, and state and local innovations led to the development of an increasingly green energy portfolio in many states and local communities. During the 1990s, the Clinton administration quietly continued to move energy policy innovation into a cooperative federal policy sphere. Post–September 11, President Bush's Hydrogen Initiative, while a long-term policy innovation, is an attempt to more rapidly move the United States into a twenty-first-century energy paradigm, although a continued need for petroleum and other carbon-based energy sources as well as a likely return to a nuclear energy paradigm is evidenced by the substance of EPAct of 2005.

NOTES

1. See Nice (1994).

2. For those eager to criticize, a note of admission; yes, clearly these categories overlap in some cases and one could easily use a partisan approach to categorizing time periods and interpretations of those time periods. The categories are fairly general, but seem to be the optimum approach to presenting an apolitical analysis of energy policy innovations.

3. It is unlikely that as free marketers, Reagan and others looked to the North Slope as a *permanent* solution to the energy crisis. Adopting principles discussed in North (1983), the free-market solution was to increase short-term energy supply and to look to energy innovations in the market place to increase energy supply for the future. In essence, the government's role should be to promote market solutions through reduced regulation and perhaps through incentive structures. Price controls were seen as artificially spurring overconsumption of a good that was in short supply, which would likely not increase the needed lead time for new market developments. North Slope oil exploration was seen as a source or energy supply that could conceivably lengthen the lead time for new innovation, while simultaneously reducing some cost

burdens on consumers. In essence, new energy supplies also reduce the inelasticity of the market and make price more sensitive to consumer demand levels.

4. The Energy and Conservation and Production Act was reauthorized in 1998.

5. A "person" could be an organization.

6. Although hydrogen-based fuel initiatives can be traced back to the Bush I administration, they were more fully operationalized under the (P.L. 104-271), which further extended legislation of six years earlier in the Spark M. Matsunaga Hydrogen Research, Development, and Demonstration Act of 1990 (P.L. 101-566). The acts were cautious yet earnest attempts to develop the use of hydrogen as fuel source as well as hydrogen fuel cell technology as a method of storing energy for transportation and other purposes.

WORKS CITED

Bobrow, D. and Kudrle, R. 1979. Energy R & D: In Tepid Pursuit of Collective Goods. *International Organization* 33(2): 149–75.

California Energy Commission. 2005. *California Energy Commission: An Overview*. www.energy.ca.gov/commission/overview.html, accessed June 11, 2005.

Clinton, W. 2000. *Executive Order 13149 Greening the Government through Federal Fleet and Transportation Efficiency*. Washington, D.C.: White House.

Cochran, T., Speth, G., and Tamplin A. 1975. A Poor Buy. *Environment* 17(4): 10–18.

Congressional Quarterly. 1992. Veto Cloud Loomed Large over 1992 Floor Fights. *Congressional Quarterly* December 19, 50(50): 3854–70.

Conway, Nicholas T. and Simay, Gregory L. 1977. Energy Research and Development: A Partnership between Federal and Local Government. *Public Administration Review* 37(6): 711–13.

Cupulos, M. 1979. The Price of Power. *National Review* 31(5): 156–59.

Economist. 1992. Does Greener Mean Richer? *Economist* 325(7788): 18–19.

———. 2005a. The Institutional Origins of the Department of Energy. Washington, D.C.: Department of Energy, www.25yearsofenergy.gov/origins.html, accessed June 13, 2005.

———. 2005b. New Stories: Factors for a Successful Renewable Portfolio Standard. *Renewable Energy & Energy Efficiency Partnership*. www.reeep.org/index.cfm ?articleid=993, accessed June 8, 2005.

Executive Order 12287. Decontrol of Crude Oil and Refined Petroleum Products, 46 F.R. 9909 (January 19, 1981).

Executive Order 13149. Greening the Government Through Federal Fleet and Transportation Efficiency, 65 F.R. 24607-24611 (April 21, 2000).

Federal Energy Regulatory Commission. 2005. *About FERC*. www.ferc.gov, accessed June 11, 2005.

Hershey, Robert D. Jr. 1981. Can Reagan Lift the Cloud over Nuclear Energy? *New York Times*, March 8, Section 3, p. 1.

Janson, Donald. 1979. Accident at Three Mile Island Nuclear Power Plant near Harrisburg, Pa., Releases Above-Normal Levels of Radiation into Atmosphere on Mar. 28. *New York Times*, March 29, p. 1.

King, Seth S. 1981. Planning of Alaska Land Use and Oil Leases Starts. *New York Times*, April 5, Section 1, p. 29.

Lazar, K. 2001. Worlds Apart. Bush, Environmentalists Battle over Earth, Wind, Fire. *The Boston Herald*, March 25, p. 1.

Munson, R. 1979. Nuclear Power: The Price Is Too High. *The Nation* 228(18): 521–39.

Neff, Shirley. 2005. *Energy Policy Act of 2005: Summary*. New York: Center for Energy, Marine Transportation, and Public Policy, Columbia University.

New York Times. 1970. February 8, p. 41.

Nice, D. 1994. *Policy Innovation in State Government*. Ames: Iowa State University Press.

North, D. 1981. *Structure and Change in Economic History*. New York: Norton.

Olatubi, W. and Zhang, Y. 2003. A Dynamic Estimation of Total Energy Demand for the Southern States. *The Review of Regional Studies* 33(2): 206–28.

Osborne, D. 1990. *Laboratories of Democracy*. Boston: Harvard Business School.

Salisbury, D. 1980. Want to Sell Your Electricity to a Utility? Go Ahead. *The Christian Science Monitor*, December 23, p. 9.

Sawyer, Stephen. 1984. State Energy Conditions and Policy Development. *Public Administration Review* 44(3): 205–14.

Schifter, I., Vera M., Diaz, L., Guzman, E. Ramos, F., and Lopez-Salinas, E. 2001. Environmental Implications on the Oxygenation of Gasoline with Ethanol in the Metropolitan Area of Mexico City. *Environmental Science and Technology* 35(24): 4957–60.

Shultz, George P. and Dam, Kenneth W. 1977. Reflections on Wage and Price Controls. *Industrial and Labor Relations Review* 30(2): 139–51.

Smith, Rodney and Phelps, Charles. 1978. The Subtle Impact of Price Controls on Domestic Oil Production. *American Economic Review* 68(2): 428–33.

Southern California Edison. 2005. *Understanding Baseline—Baseline Facts: History and Background*. www.sce.com/CustomerService/understandingBaseline/ history .htm, accessed June 11, 2005.

U.S. Congress. 1993. Energy Policy Act of 1992. 102 Congress. H.R. 776 P.L. 102-486.

U.S. Department of Energy (DOE). 2005. *The Institutional Origins of the Department of Energy*. www.25yearsofenergy.gov/origins.html, accessed June 13, 2005.

Weber, Arnold R. and Mitchell, Daniel J. 1978. Further Reflections on Wage Controls: Comment. *Industrial and Labor Relations Review* 31(2): 149–58.

WEB SITES

www.eere.energy.gov/hydrogenandfuelcells/mypp/pdfs/systems_analysis.pdf, accessed June 15, 2005.

www.ferc.gov/

Chapter Five

Solar Energy

INTRODUCTION

In the past, data related to the use of renewable/alternative energy were very difficult to collect and analyze in a consistent manner. In 1994, the Department of Energy (DOE) required the Energy Information Agency (EIA) to collect annual data on the manufacture and use of renewable energy equipment, such as solar photovoltaic (PV) panels, (see EIA 2003). Less than 1 percent of all energy consumed in the United States is produced from solar energy systems (see EIA 2004: 5). Solar energy is likely to increase its productive capacity.

WHAT IS SOLAR ENERGY? HOW DOES IT WORK?

Chapter 3 provided a basic overview of solar energy and presented an example of its use. Solar power, as concept and as a technical reality, is much more complicated. Solar power is perhaps the oldest form of energy utilized by human beings. The sun is critical to the development of agriculture-based communities and electricity used in industrial and postindustrial societies. The examples provide some sense of the division of solar energy collection and use. The former (agriculture) represents thermal solar energy, while the latter (electricity) represents the basis of electrochemical approaches.

In the 1860s, French mathematician Auguste Mouchout created the first thermal solar-powered steam engine. Mouchout, and later Englishman William Adams, devised methods by which solar radiation could be concentrated on tanks of water. Solar radiation would heat the water, creating steam power. In the 1870s, American John Ericsson created the first solar trough, which was a long parabolic device that focuses solar radiation onto a pipe carrying an efficient thermal-absorbing liquid, usually water or oil, that would absorb and transport solar thermal energy to turbines used in generating electrical energy. Solar thermal power continued to evolve as the primary form of solar power research until the 1910s (Smith 1995 [www.solarenergy.com/info_history.html, accessed June 17, 2005]).

Electrochemical applications of solar energy was pioneered by French physicist Alexandre Bequerel (Lenardic 2005). In thermal solar processes, capturing energy from light seems rather simple—the warmth of the sun heats up a liquid that, in early applications, can be converted into steam and then used to power steam engines. In electrochemical processes, however, the energy contained in light must be directly captured in the form of electrical energy. Becquerel and other late nineteenth-century scientists observed the impact of illumination on electrochemical phenomena, but it was not until Einstein's work in 1905 that the PV effect was more fully understood. Einstein theorized about the dual nature of light as both a particle and a wave. The particle aspect of light is known as a photon—a small but powerful bundle of energy.

It is one thing to use illumination to change the nature of an electrochemical process; it is quite another to make photons create a controlled electrical energy. PV cells absorb solar radiation—the energy of photons—and direct the movement of electrons in a controlled pattern to produce electrical voltage and current (i.e., electrical power). The material that absorbs photons in a PV cell is a semiconductor, generally made of silicon.

Of particular importance to PV systems is the reduction of unwanted impurities in the silicon material and to create a uniform crystal structure. Silicon is a highly desirable material because it can be manufactured into a controlled crystalline form. Monocrystalline silicon is a premium material because the crystal formation is uniform and free of impurities, which makes the substance very efficient at absorbing energy from photons, while polycrystalline silicon has a lower efficiency for absorbing energy from photons. Silicon can be cut into very thin translucent wafers that effectively absorb photons. Traditionally, a standard PV cell is approximately 300 microns thick (Green 2000: 348).[1]

Manufacturers introduce a very small amount of impurity into the silicon crystal so the silicon can build a charge while irradiated by solar energy. N Type picks up extra electrons and is negatively charged. Silicon can also be charged "positive" (P Type), which means that the materials need extra electrons. N-Type silicon has phosphorus added to it, while P-Type silicon has

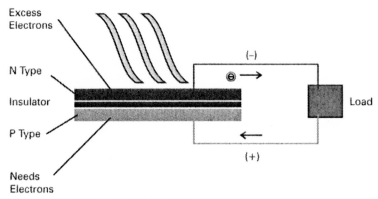

Figure 5.1. Schematic of Solar Cell

boron added.[2] The process of making the silicon N or P Type is called doping. P-type and N-type silicon are placed adjacent to one another but separated by a silicon oxide insulating layer, which prevents electrons from moving directly between the two silicon media. As a solar energy system, the silicon cell now has a charge and is similar to a battery (see figure 5.1). The electrons are then forced to travel through an electrical circuit, which directs the electrons through some form of electrical mechanical device, such as a water pump, that uses the electrical energy. Electrical energy can also be stored in a battery for latter use.

PV systems using monocrystalline silicon were first developed in the 1940s, but the cost of producing the PV cells was prohibitive. In the 1950s and 1960s, PV technology was used in the space industry as a form of energy to power satellites and other outer space-related system applications.

Solar thermal energy remains an important part of solar power technological development and use, particularly in residential and small commercial applications. The load requirements for well-designed energy-efficient homes are much lower because of hi-tech insulation, low-energy requirement equipment such as high-efficiency hot water heaters, dishwashers, and laundry machines. Additionally, proper site location and structural consideration can make solar energy a better choice simply because of reduced electric lighting requirements and the solar thermal benefits from more efficient home design.

Several technical and economic issues are addressed in the following sections, which provide important information about the current feasibility of solar energy and its future growth potential, particularly electrochemical processes. Understanding current technical and economic issues will produce better energy choices that properly incorporate solar energy systems.

TECHNICAL FEASIBILITY OF SOLAR POWER

Technical issues often relate to the development of efficient and effective energy collection devices for PV cells and cell housing. As shown in figure 5.1, it is necessary to have wires connected to solar cells to collect electrons that are freed from the N-Type material due to exposure to photons emitted by the sun. Although silicon has many benefits, one of its great strengths is simultaneously one of its greatest weaknesses; namely, its status as a semiconductive material. Semiconductors are not very effective in creating an electrical current because their atomic structure is such that their electrons are fairly stable. Electrons trying to move through the outer shell of the silicon atom are not really able to do that very effectively. Therefore, a metal grid of collectors is constructed to overlay the PV cell to collect electrons.

Collector grids block cell exposure to photons, which means that a small portion of the cell is not producing electrical current. Even a small amount of shading can significantly reduce cell efficiency. Collector grids are composed of highly conductive metal able to effectively transport electrons through very thin wires yet retain its structural integrity and reduce shading. In order to meet the necessary standards of functionality, metal conductor grids require expensive materials and manufacturing processes. Other alternatives to dealing with metal grid impacts on cell efficiencies include imbedding the metal grid in grooves along the cell surface. A technique known as photolithography is used to literally paint the metal grid onto the surface of PV cells. An optimum material used for collector grids is tin oxide (SnO_2). The advantage to certain grid materials is that they are nearly transparent and do not shade the surface of the silicon materials where electrons are being freed from N-type material, yet the materials retain integrity despite exposure to temperature extremes and solar radiation patterns (see Energy Efficiency and Renewable Energy 2005 [www.eere.energy.gov/solar/electrical_contacts.html, accessed June 17, 2005]).

Nanotechnology is being investigated in the hopes of constructing cheaper and more efficient PV cells. The components of these efficient cells are so small that it is necessary to construct the cells on a nearly atom-by-atom basis, which is a hallmark of the nanotechnology revolution in high-tech manufacturing. As mentioned previously, photolithography is an important part of modern solar cell construction, but nanotechnology also promises improved solar efficiencies. Since light is composed of many different frequencies, it is necessary to efficiently capture photons over the entire spectrum. These frequency differentials require different N-type materials within the solar cell, but will also require newly emerging techniques of making these spectral frequencies into useful electrical energy for different system load requirements.

Thin film PV technology has been a continually evolving concept since PV systems were developed in earnest in the mid-twentieth century. The concept behind a thin film system is to create ultra-thin cell compositions (e.g., collector components of ≤150–200 microns in thickness). Thinner solar cells have been produced that claim to have higher efficiency ratings (see Day Star Technologies 2004), which means that the cells effectively capture a higher percentage of the energy found in the sun's radiation. Efficiency ratings cited by Green (2000: 449) for thin film technologies were approximately 18 percent, consistent with earlier evidence from studies conducted at Sandia Laboratories using 15-micron-thick thin cells produced efficiency ratings of 17.6 percent (see Zheng et al. 1998). Thin film technologies remain in development stages because of the impact of solar radiation degradation on materials within the systems (see Han 1998), but degradation is dependent on the material composition of the solar cell. Thin film production poses material safety issues (Green 2000: 449), as some cell compositions employ environmentally "toxic" materials, such cadmium telluride.

Another problem that arises from thin film technologies relates to durability. Solar systems are exposed to high levels of solar radiation, high temperatures, wind, dust, water, and other contaminants or corrosives. Due to the very thin structure of thin film technologies and the efforts to reduce glass protective plating requirements.[3] The exposure of thin films makes the product much more likely to delaminate and to fail in operation (see McMahon 2004). Technical research has focused on combining solar thermal and PV systems to, in essence, cool solar panels, thus increasing efficiencies while simultaneously reducing the negative impact of intense heat on materials and component integrity (see Tripanagnostopoulos et al. 2005).[4]

Efficiency ratings for solar systems improved quite steadily over the 1990s, and have continued to improve in the early twenty-first century. Efficiency rating improvements are, in large measure, a function of continued research and development in PVs and solar thermal systems. Monocrystalline silicon cells have produced documented efficiency ratings of nearly 25 percent while polycrystalline and thin film studies have produced mean efficiency ratings of 20.3 percent and 17.6 percent, respectively. Different studies reported by Green et al. (2005) produce slightly different mean efficiency results, but in nearly all cases differences are within the margin of error (see Green et al. 2005). The prospect for continued improvement in PV cell efficiency remains viable, but PV researchers expect efficiencies to reach ~30 percent as a maximum efficiency; although Kazmerski and Broussard (2004) claimed one year earlier that rates have reached nearly 40 percent and may move toward even greater efficiency. Material issues and efficiency ratings in PV cells will likely have a significant impact on the economics of solar power, increasing its role in producing energy supplies for the twenty-first century.

ECONOMICS OF SOLAR POWER

The economics of solar power can be divided into many subtopics. First, what type of economic infrastructure currently exists for solar power? What are the historic costs of solar energy systems and in which direction are costs trending? Second, solar power should be studied to determine its feasibility as a replacement energy source, reducing the need for carbon-based energy sources. How does $/kWh (kilowatt hour) for solar energy compare with carbon-based energy sources, and what are the future prospects for $/kWh? Also, how does solar energy as a replacement reduce economic and social costs associated with carbon-based energy systems? A third question relates to the economic development opportunities associated with the development of a solar energy paradigm. Do solar energy systems create job opportunities? These questions become quite important when one considers the sustainable community model undergirding much of the alternative/renewable energy paradigm. A sustainable community involves a great deal more than a clean environment built on a sustainable possibly "green" energy paradigm—citizens also need an economic base providing them with income and products and services to be purchased or sold.

Current Solar Energy Economic Infrastructure

The EIA of the DOE is a recognized data source regarding energy production and consumption in the United States. Its most recent publicized data indicate that in 2005, U.S energy consumption was approximately 100 quadrillion British thermal units (BTUs);[5] according to the EIA report, approximately 63 trillion BTUs, less than 1 percent of the energy consumed was created using solar power (EIA 2005). "Solar PVs have expanded rapidly in recent years, but their share of the [total energy production and consumption] is so small that this growth has not affected the renewable industry trend significantly" (EIA 2004: 1). For use in the production of commercial electricity, solar energy consumption has not changed significantly in terms of consumption patterns since 1999 (EIA 2005). The use of PVs expanded during the period of study, but solar thermal systems declined in use. The net electricity generation from solar power increased from 495.1 MWh (megawatt hour) in 1999 to 579.0 MWh in 2004—an increase of approximately 17 percent. During the six-year period of study (1999–2004), solar energy capacity increased by 0.2 percent, while total nonrenewable energy capacity increased by 1.4 percent and nonrenewable energy capacity increased by 25 percent. This finding is particularly critical in studying the economic feasibility of solar PVs as a potential replacement energy source for carbon-based fuels. At this time, solar energy remains underutilized *if* solar energy is widely considered a potential replacement energy source for the future.

Generally, solar energy production and consumption occurs in high solar radiation areas of the nation. Climatalogically, the southern Pacific coast area and the inland southwestern United States are prime locations for solar energy systems. Solar energy is used for electricity production, efficiently collected in Arizona and California; otherwise, EIA reports near zero net electricity production from solar elsewhere in the nation (EIA 2004: 15). Residential use of solar PV and thermal systems is more widespread and largely dependent on solar radiation quality. State and federal incentives for the use of alternative/renewable energy systems play a substantial role in the economic choice to purchase and utilize residential solar energy systems, which also impacts independent power producers seeking economic benefits from selling zero emission tax credits.

In 2002, solar energy was approximately 13 percent of all renewable energy consumed in households (EIA 2004: 6), compared with 0.1 percent of electricity power generation by utilities and independent power generators. Household production and use of solar electricity was similarly about 0.1 percent, which would indicate that solar power was used for nonelectricity purposes.

The evidence indicates that solar power is very modestly enlarging its share of the energy supply in the United States. One energy-producing sector that is showing a sizable increase in solar energy use is an approximate 67 percent growth in energy production by independent power producers between 1989 and 2003 (from 3 trillion BTUs to 5 trillion BTUs; for comparison, in 2003 total renewable energy consumption was 6,131 trillion BTUs). Paralleling this increase in solar energy production by independent power producers, there is evidence that solar system manufacturing and purchases are increasing.

Nearly 15 million square feet of solar thermal collectors were shipped by domestic manufacturers to consumers in 2005. Approximately 30 thousand peak kWh solar cells and 134.5 thousand peak kWh PV modules were shipped by domestic manufacturers in 2005. Peak kWh in this case means that the number of PV cells and modules shipped had the capacity to produce over 220 thousand kW on an hourly basis during peak solar PV performance, under optimum solar radiation conditions.

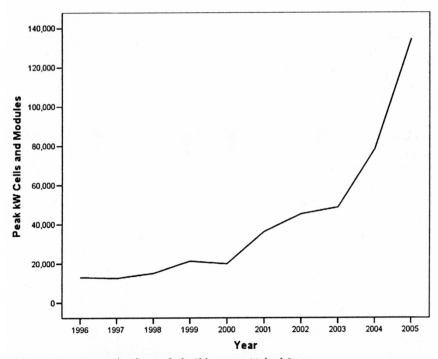

Figure 5.2.   Domestic Photovoltaic Shipments, United States
Source: Energy Information Agency 2003. *Renewable Energy Annual 2003*, Solar Thermal and Photovoltaic Manufacturing Activities, Table 5. Washington, D.C.: U.S. Department of Energy, www.eia.doe
.gov/cneaf/solar.renewables/ page/solarreport/solar.html, accessed October 4, 2006.

Figure 5.2 indicates that shipments of PV cells and modules have increased
dramatically since 1994. Domestic demand shot up in 2000, while export demand has dropped slightly since 2000. While demand for solar PV cells has
increased, prices for solar modules and cells has decreased since 1994,
which is partially a function of the technical advances discussed previously.
PV systems are also cheaper to build because of the increased capacity to
mass produce the technology.

Economic Development Impacts

As stated previously, the economic development impacts of adopting solar
energy are of particular interest to individuals developing sustainable communities. Several economic analyses of solar energy have shown a net positive economic impact of solar energy, particularly the large-scale adoption of
solar energy in communities. One method of looking at economic development impacts is to study the economic outflow associated with importing en-

ergy resources into a community. The Sacramento Municipal Utility District found that an economic multiplier effect of 2.11 resulted from its solar energy and energy efficiency programs, which reduced energy imports (DOE 1996). A comparative analysis completed by Schwer and Riddel (2004) presented three solar energy plant construction and operations and maintenance options ranging from 100 MW- to 1,000 MW-size solar energy systems. Schwer and Riddel (2004) estimate an employment multiplier effect between 2.9 and 3.9 and gross state product would be increased between $1.15 billion and $3.5 billion over an assumed thirty-year lifetime of the project.

Solar energy systems, particularly large-scale systems discussed here, demonstrate the real economic potential for states and local governments to build sustainable communities. Schwer and Riddel (2004) make the point that green energy is an additional benefit, as increasing renewable energy is part of Nevada's mandated energy portfolio. Thus, there is a multidimensional benefit to developing large-scale solar energy systems. While Nevada is only one example, several other states have renewable energy requirements and would also benefit—economically and otherwise—from a strong commitment to renewable energy systems such as solar (see also Campbell and Pape 1999).

FEDERAL SOLAR ENERGY TECHNOLOGIES PROGRAM

With fossil energy still relatively abundant and prices within the reach of most consumers, it is unlikely that solar energy will become prevalent for several years. In fact, if other energy options are adopted that are cheaper and are technically and political feasible, then solar may remain in the shadows of the marketplace for several decades. That being said, solar energy is politically acceptable and increasingly technically and economically possible largely because of the commitment of policy leaders, private consumers, and industry to the notion of clean and safe energy for sustainable communities. At some point, a replacement energy source must be found to match the energy demands currently placed on hydrocarbon sources. It is better to have multiple supply options available rather than only a handful, and it is better to have well-developed options rather than unproven alternatives. In the postindustrial Third Way political and economic environment, it seemingly falls on government *and* the private sector to pursue a seamless transition to a new energy paradigm, monitoring energy needs in relation to energy supply options. In the federal system of government found in the United States, federal, state, and local efforts are required to coordinate public policy—in this case, alternative/renewable energy policy.

At the federal level, one program seeking to promote continued research into ever-more efficient solar energy systems is the Solar Energy Technology Program (SETP). The program's goals are as follows:

* Improve the cost, integration, and performance of solar heat, cooling, electricity, and lighting technologies in combination with building systems to levels where they are a competitive, reliable option for building owners and occupants.
* Add significant security, reliability, and diversity to the U.S. energy system and improve the quality of life in this country by expanding inexpensive electricity to all.
* Make solar technologies and systems an accepted and easily integrated option for distributed-energy production—both on and off the electric utility grid.
* Develop next-generation technologies and systems with the potential to create new high-value applications of solar energy in producing hydrogen fuel, generating competitive bulk power at central stations, desalinating water, or creating other products that are beyond present capabilities.
* Reduce the environmental signature (air emissions) by displacing fossil-fuel energy systems with cost-effective solar energy systems.

Quoted from: Energy Efficiency and Renewable Energy 2005.

According to the program's analysis, solar energy costs have declined significantly, but still remain above the cost of hydrocarbon energy sources. Projections indicated that cost will remain either slightly higher than or at parity with carbon-based fuel costs for the remainder of the decade, assuming that current carbon-based electricity generation systems maintain current prices.

SETP seeks to streamline market processes to advance solar energy potential in the near future (SETP 2005d). Program leaders seek to use a systematic approach to research and development in the solar industry and to promote the use of solar energy by commercial enterprises as a form of supplemental energy for factories and offices. The program also seeks to coordinate and/or monitor the solar industry in various manufacturing and retail subsectors ranging from "materials" and "component" manufacturing to "applications" and "markets" (see www.eere.energy.gov/solar/ systems_driven .html, accessed June 23, 2005]). Universities and private industry partnerships with SETP are encouraged through grant opportunities and collaborative research and development.

Through the Financing Solutions and Incentives office within the program's parent organization, the Energy Efficiency and Renewable Energy Office, DOE, homeowners, private industries, utilities, and government organizations are offered resources or financial incentives to promote the use of solar energy. The Financial Solutions program provides access to government incentives for residential and commercial energy consumers that choose to adopt solar energy systems in their work environment or homes. Initiatives such as the Million Solar Roofs Initiative and the Utility Solar Water Heating Initiative represent two significant policies promoted by SETP (see Solar Energy Technologies Program 2005b [www.eere.energy.gov/solar/deployment .html, accessed June 23, 2005]). The Million Solar Roofs Initiative counts 350,000 solar roof installations since 1997 toward its goal of 1 million installations by 2010. Solar heating and lighting initiatives are also promoted by SETP.

The program budget for the last year was approximately $83.4 million for federal year (FY) 2004, which is less than a 2 percent increase over FY 2003. In real dollar terms, the budget change represents a decline in spending power. In terms of national political priorities, this would indicate that SETP is not a high priority in national energy policy; at the very least, one could conclude that it is not a growing priority area. The vast majority—approximately 90 percent—of the SETP budget is devoted to promoting solar PVs, while the remainder of the budget focuses on concentrating solar power and solar heating/lighting programs (see Solar Energy Technologies Program 2005a [www.eere .energy.gov/solar/budget.html, accessed June 23, 2005]).

From an economic theory perspective, SETP is an example of Third Way politics at work; namely, public/private cooperative relationships promoting pragmatic solutions to solve a policy problem. From a neo-institutional theory perspective, the sovereign (in the form of SETP, in this instance) seeks to structure the decision environment through incentives and regulations.

STATE AND LOCAL EFFORTS

State and local efforts focus on renewable/alternative energy emphasizing renewable energy portfolio standards, which establish green energy targets, and through loan and tax incentive-based programs, focusing specifically on solar energy use. Renewable Portfolio Standards (RPSs) shape markets by creating new incentives and constraints, ultimately shaping consumer behavior in the energy market. Tax incentive policies steer policymakers and citizens toward new types of energy purchase decisions.

The state of California leads the nation in purchasing of solar energy systems. Property tax incentives are offered to energy producers. Under the California Revenue and Taxation Code, Section 73, property tax assessments are not levied on approved solar energy systems. Additionally, the state offers Supplemental Energy Payments (SEPs) to alternative/renewable energy producers who receive below market prices for energy sales to utilities; the SEP brings total energy payments up to market prices for energy sold to consumers. Washington State uses a slightly different incentive program for solar power generation, providing a 15¢/kWh production incentive, up to $2,000 (see RCW 82.16, SB5101, Washington State Senate, 2005). Ferry County in Washington State has offered 0 percent interest loans to individuals purchasing and installing solar energy systems for their homes. Ferry County is a rural county and has a very high rate of citizens living in poverty. The solar energy incentive helps those citizens who are paying more than 23¢/kWh for electric energy. Conservation and Renewable Energy Discounts are rebates for individuals installing solar energy systems in Klickitat Public Utilities District. The rebate provides $400 per kW of solar energy panels up to 3kW. Similar programs are available in several other counties in the state.

Oregon provides similar incentive programs; as is the case in Washington State, the Bonneville Environmental Foundation (BEF, associated with the Bonneville Power Administration) provides a 10¢/kWh incentive for solar energy through something known as Green Tag purchasing. A nonprofit organization called The Northwest Solar Co-Op sells Green Tags to solar energy producers, which certify how much green energy is being produced by the systems. The BEF (a nonprofit firm associated with but independent of Bonneville Power Administration) buys Green Tags, which is, in essence, a form of emissions trading. The BEF is effectively reducing their emissions by offsetting commercial power generation emissions with the green benefits purchased by the BEF from green energy producers (Cascade Solar Consulting 2005 [www.cascadesolar.com/ssguide.pdf, accessed June 23, 2005]). BEF is also actively involved in promoting solar powered schools through its 4R Schools program; a public-private nonprofit partnership (BEF 2005b [www.b-e-f.org/grants/solar.shtm, June 23, 2005]). State and local incentives and tax exemptions for solar power producers and users are also used to promote solar energy.

Alaska offers very little in the way of incentive programs for solar energy production, perhaps due to the state's tremendous fossil energy resources. Hawaii, however, with no available fossil energy resources, has a fairly well-developed system of tax incentive programs for residential solar thermal and solar energy producers, such as the state Residential Solar and Wind Energy Credit (Hawaii SB3162 of 2004), Maui Electric Company solar water-heating program (Maui Electric Company 2005 [www.mauielectric.com/MECO/page/, accessed June 23, 2005]), and the Honolulu Solar Roofs Initiative offering 0 percent to 2 percent low interest loans to residential solar water heating system purchases (Honolulu Electric Company 2005 [www.heco.com, accessed June 23, 2005]).

Why do state and local governments need to use grassroots innovation to promote solar energy? Perhaps one Hawaiian electric company's website provides some clues to the impending energy crisis of the early twenty-first century. The Kaui'i Electric Company website demonstrates the need to promote solar energy as an alternative, given that electric energy costs have increased from less than 7¢/kWh in 2003 to over 15¢/kWh in 2005. National programs are one way of developing solar energy policy innovations, focusing primary attention on macroeconomic issues that will transform industry and the U.S. economy in a way that is more consistent with a renewable energy paradigm; but, the state and local initiatives show a tremendous effort to promote microeconomic changes in consumer and producer behavior. With the aid of grants and tax incentives, the energy consumer is getting used to becoming the energy producer, too! Although covering only West Coast states, the analysis does provide several examples of states and local governments working in conjunction with nonprofit and for-profit enterprises to promote the use of solar energy. Consumer demand for solar energy technology is rising, but remains a small portion of the overall energy portfolio. Demand patterns are indicative of a need for greater citizen support for solar energy if it is to play a larger role in energy generation.

CHAPTER SUMMARY

Solar energy policy relies heavily on consumer awareness of a product, its benefits, and its costs. Despite citizen/consumer benefit structures that have been put in place through public policy innovations, citizens' energy choices are increasingly moving in the direction of solar energy. Full understanding of the increasingly seamless technological interface of solar energy systems and residential and commercial energy demand may not be understood by potential consumers. The technology behind solar energy policy is solid and continues to improve as efficiency ratings approach 40 percent, which means smaller systems effectively meeting load demands. The economic behind solar energy also remains quite solid with national, state, and local government incentives as well as a substantial domestic solar energy system production infrastructure. As fossil energy prices continue to rise, citizens/consumers are likely to increasingly find that costs for solar energy alternatives may be a viable solution for their energy needs and a significant way in which individuals can contribute to community sustainability.

NOTES

1. Professor Yang Yang and colleagues at Henry Samueli School of Engineering and Applied Science at University of California, Los Angeles, have conducted noteworthy research on the feasibility of a plastic solar cell. Still in development phase, the plastic solar cell has an efficiency rating of 4.4 percent, compared to silicon cells that have ratings between 14 and 18 percent (Abraham 2005).

2. The first PVs developed at Bell Laboratories used boron substrates (see Green 2000: 443).

3. In a typical comparative solar cell study, Fanney et al. (2002: 2) employed 6mm glass plating, which reduces collector efficiencies but protects solar cells from damage. Fanney et al. found that the conversion efficiency controlling for cell area was highest for monocrystalline cells, ranging from 10.5 percent to 12.5 percent. Thin film cell efficiency ranged from 4.8 percent to 6.9 percent. The efficiencies reported illustrate the variations in cell efficiencies probably related to the methodology of study and local conditions of study. Amphorous silicon can be used in thin film cells. Unlike crystalline formation, amphorous silicon can be produced by super heating silicon and then rapidly cooling in thin sheets. Crystalline formations are more likely to establish collector points based on uniformity of doped silicon, but crystalline formations—particularly monocrystalline silicon—is more expensive to produce. Material development and use is, at least partially, a function of economic trade-offs.

4. Other types of materials are being actively explored by solar technology researchers. One type of material that will be of future interest to manufacturers and users of solar energy is found in the plants that surround us: chlorophyll. Material scientists explore the value of chlorophyll as a material for collecting electricity in a solar cell and have produced evidence of success. Yun et al. (2005) created a chlorophyll-based solar cell with a conversion efficiency of 1.48 percent. As a material, chlorophyll is abundant. With continued technical development, it could be a technically efficient and effective method of collecting energy for meeting electrical energy load demands. The technical achievement is consistent with other research being conducted in materials science, applied physics, and electrical engineering that have successfully advanced the solar energy paradigm beyond crystalline silicon.

5. To provide some perspective, one BTU is the amount of energy required to raise one pound of water 1° F. Dry wood has 7,000 BTUs of energy per pound (see Riches 2005 [bbq.about.com/od/gasgrills/g/gbtu.htm, accessed June 21, 2005]).

WORKS CITED

Abraham, M. 2005. Engineers Pioneer Affordable Alternative Energy Resources—Solar Cells Made of Everyday Plastic. *News Center: Henry Samueli School of Engineering and Applied Science, UCLA.* October 10. www.engineer.ucla.edu/news/2005/plasticsolarcells.html, accessed March 11, 2006.

Bonneville Environmental Foundation (BEF). 2005a. *Green Tags.* www.greentagsusa .org/GreenTags/index.cfm, accessed June 23, 2005.

Bonneville Environmental Foundation (BEF). 2005b. Solar Projects for Homes and Schools. www.b-e-f.org/grants/solar.shtm, accessed June 23, 2005.

Campbell, B. and Pape, A. 1999. *Economic Development from Renewable Energy: Yukon Opportunities.* Drayton Valley, Alberta, Canada: The Pembina Institute.

Cascade Solar Consulting. 2005. *Solar Starter Guide.* www.cascadesolar.com/ssguide .pdf, accessed June 23, 2005.

Day Star Technologies. 2004. *Products.* www.abanet.org/media/faqjury.html, accessed June 20, 2005.

Energy Efficiency and Renewable Energy. 2005. *Electrical Contracts.* www.eere .energy.gov/solar/electrical_contacts.html, accessed June 17, 2005.

Energy Information Agency (EIA). 2003. *Renewable Energy Annual.* Washington, D.C.: U.S. Department of Energy. www.eia.doe.gov/cneaf/solar.renewables/page/ solarreport/solar.html, accessed October 4, 2006.

Energy Information Agency (EIA). 2004. *Renewable Energy Trends (with Preliminary Data for 2003).* Washington, D.C.: U.S. Department of Energy. www .eia.doe.gov/cneaf/solar.renewables/page/trends/trends.pdf, accessed June 17, 2005.

Fanney, A., Dougherty, B., and Davis, M. 2002. Performance and Characterization of Building Integrated Photovoltaic Panels. *Proceedings of the 28th Annual IEEE Photovoltaic Specialists Conference.* May 20–24, New Orleans.

Green, Martin A. 2000. Silicon Solar Cells: At the Crossroads. *Progress in Photovoltaics: Research and Applications* 8: 443–50.

Green, Martin A., Emery, Keith, King, David L., Sanekazu, Igari and Warta, Wilhelm. 2005. Solar Cell Efficiency Tables (Version 25). *Progress in Photovoltaics: Research and Applications* 13, 49–54.

Han, D. 1998. *Experimental Study of the Factors Governing the Staebler-Wronski Photodegradation Effect in a-Si: H Solar Cells.* Golden, CO: National Renewable Energy Laboratories.

Honolulu Electric Company. 2005. *Honolulu Solar Roofs Initiative Loan Program.* www.heco.com, accessed June 23, 2005.

Kazmerski, L. and Broussard, K. 2004. *Solar Photovoltaic Hydrogen: The Technologies and Their Place in Our Roadmaps and Energy Economics.* Prepared for the *19th Annual European PV Solar Energy Conference and Exhibition.* Golden, CO: National Renewable Energy Laboratory, August. NREL/CP-520-36401.

Lenardic, Denis 2005. A Walk through Time. *PV Resources.com.* www.pvresources .com/en/history.php, accessed June 18, 2005.

Maui Electric Company. 2005. *Welcome to Maui Electric!* www.mauielectric.com/ MECO/page/, accessed June 23, 2005.

McMahon, T.J. 2004. Accelerated Testing and Failure of Thin-Film PV Modules. *Progress in Photovoltaics: Research and Applications* 12, 235–48.

Riches, Derrick 2005. *BTU.* http://bbq.about.com/od/gasgrills/g/gbtu.htm, accessed June 21, 2005.

Schwer, R. and Riddel, M. 2004. *The Potential Economic Impact of Constructing and Operating Solar Power Generation Facilities in Nevada.* NREL/SR-550-35037. Golden, CO: National Renewable Energy Laboratories.

Smith, Charles. 1995. *History of Solar Energy: Revisiting Solar Power's Past.* www .solarenergy.com/info_history.html, accessed June 17, 2005 [see also Smith 1995, Solar Power in *Technology Review*].

Solar Energy Technologies Program. 2005a. *Budget.* www.eere.energy.gov/solar/ budget.html, accessed June 23, 2005.

Solar Energy Technologies Program. 2005b. *Deployment.* www.eere.energy.gov/ solar/deployment.html, accessed June 23, 2005.

Solar Energy Technologies Program. 2005c. *Implementing a Systems-Driven Approach.* www.eere.energy.gov/solar/systems_driven.html, accessed June 23, 2005.

Solar Energy Technologies Program 2005d. Mission, Vision, Goals. www.eere .energy.gov/solar/mission_vision_goals.html, accessed June 23, 2005.

Tripanagnostopoulos, Y., Soutliotis, M, Battisti, R., and Corrado, A. 2005. Energy, Cost and LCA Results of PV and Hybrid PV/T Solar Systems. *Progress in Photovoltaics: Research and Applications* 13: 235–50.

U.S. Department of Energy (DOE). 1996. The Jobs Connection: Energy Use and Local Economic Development. *Tomorrow's Energy Today for Cities and Counties.* Washington, D.C.: U.S. Department of Energy.

Yun, J., Jung, H., Kim, S., Han, E., Vaithianathan, V., and Jenekhe, S. 2005. Chlorophyll-Layer-Inserted Poly (3-Hexyl-Thiophene) Solar Cell Having a High Light-to-Current Conversion Efficiency of up to 1.48 Percent. *Applied Physics Letters* 87(2): 12–14.

Zheng, Guang Fu, Wenham, Stuart R. and Green, Martin A. 1998. Short Communication: 17.6 Percent Efficient Multilayer Thin-Film Silicon Solar Cells Deposited on Heavily Doped Silicon Substrates. *Progress in Photovoltaics: Research and Applications* 4(5): 369–73.

Chapter Six

Wind Energy

INTRODUCTION

Wind energy is one of the oldest forms of energy generation. The growth of fossil fuel economies in the late nineteenth and early twentieth centuries led many energy consumers to view wind energy systems as inferior. Fossil energy was readily available and accessible on demand while wind systems were dependent on weather conditions. In a new era, hydrocarbon energy sources are beginning to show the strain of increasing global demand; costs and supply-related issues are making reliance on carbon-based energy sources less certain. Rising costs for energy combined with technological developments and government incentives that have effectively reduced direct costs to energy consumers are making wind energy an increasingly viable alternative to carbon-based energy sources.

Wind energy is one of the fastest-growing renewable energy sources in the United States. The cost of wind energy is declining in large part due to improved siting technology, wind blade control and structure, and improved wind generator systems. The oil shortages in the 1970s and early 1980s led to tremendous growth in research and development funding for wind energy in the United States, as suppliers and consumers sought an expansion of the energy supply base. While U.S. energy policy operated on a reduced commitment to alternative or renewable energy in the 1980s and 1990s, many Western European nations continued their development of wind energy technology. In the post–September 11 world of energy uncertainty and as petroleum prices rose to reflect growing global energy demand, the United States has renewed its commitment to the development of alternative and/or renewable energy systems.

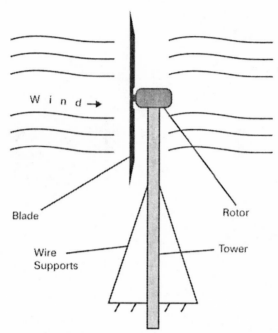

Figure 6.1.   Wind Turbine Construction
Drawn by Christopher A. Simon.

WHAT IS WIND POWER? HOW DOES IT WORK?

Wind power technology is one of the oldest forms of energy generation in human civilizations. Early mechanical systems were used nearly seven thousand years ago in Egypt and the Middle East to operate windmills for grain processing. In the United States, windmills were used well into the twentieth century, but were rapidly replaced by steam and electrical power. Wind-powered water pumps were widely used, particularly in rural applications, in the United States for much of the nineteenth century and for several decades in the twentieth. The Rural Electrification Act of 1936 (7 USC 31) provided cheap and abundant hydropower electricity to many rural areas, making wind generation an inferior energy generation system for the modern age (see figure 6.1 and figure 6.2).

Wind power systems have at least three parts: propeller blades, rotors, and support towers. Blades for wind power systems use many of the same principles as are used to construct blades for propeller-driven aircraft and for engine propellers used on ships.

Important parts of a propeller:

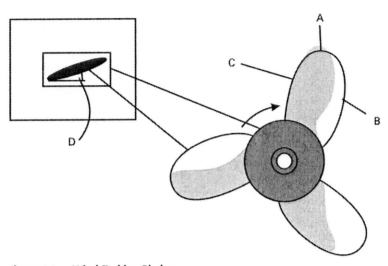

Figure 6.2. Wind Turbine Blades
Blade assembly portion of rotor drawn by Christopher A. Simon.

- blade tip (A): the very end of a propeller blade. Blade tip to center of rotor is one way to measure blade length. While the entire blade may be revolving twenty times per minute, the blade tip, due to the length of the blade, might be moving at 150 mph.
- leading edge (B): the part of the blade with which wind first comes into contact.
- Trailing edge (C): the part of the blade with which wind is last in contact.
- Pitch (D): the angle at which the blade sits, as measured by the imaginary perpendicular line in relation to the wind. Blade pitch depends on the strength of the wind in sustained wind speed as well in terms of wind gust speed. When the angle (D) becomes larger, the pitch of the blade is considered deeper, which is typical in wind systems operating in areas with high sustained winds and/or high-speed wind gusts. When the angle (D) is smaller, then the pitch of the blade is considered more shallow, which is often the case when sustained wind speeds are much slower and/or wind gusts are much lighter.

The rotor is the central feature of the blade assembly. At the rotor juncture, blades are attached to the wind power system. Within the rotor assembly, at least two different control and power take-off processes occur. First, blade pitch control exists within the rotor unit, allowing wind power user to adjust pitch depending on wind speed. At times, pitch control is automatically controlled by the wind power unit itself. Second, the rotor assembly contains a system of gears that mechanically increase the rotation speed of the electrical generator or other power take-off system—the nacelle. The nacelle also houses a generator for the production of electricity. Due to the system of gears, rotation speeds within the generator unit may exceed 1,500 rpms (revolutions per minutes) (see Energy Center of Wisconsin 2005). Within the nacelle, generated power fed onto the electrical grid must match the 60 hertz (or cycles per second) of conventional power systems in the United States. In contemporary wind power generation, power take-off is usually in the form of alternating current electricity. Even in light winds, properly designed propeller blades, often constructed from strong lightweight materials (e.g., aluminum or fiberglass), maintained at an optimum pitch, are capable of generating significant amounts of electricity.

Wind energy towers are particularly important in the energy generation process. Tower height is often a function of the length of propeller blades and size of rotor assembly. Additionally, wind speed analyses at various points above ground level may reveal optimum wind speed conditions; tower height and propeller blade length have to be properly sized and pitched so that efficient wind speeds are captured by wind generation systems, while simultaneously minimizing the impact of wind velocity and turbulence on tower and other wind turbine physical plant harmonics and load (see Larsen et al. 2005). Modern wind turbine towers are typically between 150 and 200 feet tall and 10 feet in diameter. Towers are either solid metal structures or are a lattice design. Towers are anchored into the ground and the strength of the anchoring depends on tower size and wind speed at generating site. Tower footings may be as much thirty feet deep. Cable support wires may be used to assist in supporting the tower and prevent torsion forces from causing damaging sway motion.

TECHNICAL FEASIBILITY OF WIND POWER SYSTEMS

Wind power is becoming increasingly feasible because of tremendous technological developments. Technological developments discussed in this section will relate to materials-related achievements that have increased the attractiveness of wind energy systems in a variety of locations. Additionally, the chapter will discuss location decisions that are becoming better informed because of wind-tracking data systems. Materials, technology, and informed decision making are important to nearly any power generation process.

Material accomplishments in wind power have impacted all aspects of the basic wind energy framework discussed in the previous section. Propeller technology continues to change because of material developments and generation system refinements related to the wider use of wind energy globally. Different locations require that energy systems meet different conditions and needs. For instance, in locations with very strong winds or particularly high wind gust regions, systems must be designed to resist propeller and tower damage, while simultaneously maintaining high levels of efficiency and properly meeting load requirements. Using a lightweight propeller might produce greater power; but in a high wind gust area it could be severely damaged. Material fatigue is another factor that could increase maintenance costs. Due to technological developments, wind energy systems can be footed in the ocean with propellers and rotor assemblies placed just above ocean level. In an ocean or sea air environment, proper materials must be used to resist the corrosive impact of salt water. Finally, low wind environments require the use of materials that will more easily capture the force of the wind and produce usable electrical energy.

Early propeller systems were constructed from wood or lightweight metals, such as aluminum. In some instances, aircraft propellers were used in wind energy systems. Wooden propellers pose many problems, such as cracking as well as material decomposition due to moisture, particulate matter, and other environmental conditions. Additionally, wooden propellers may be heavy, thus requiring stronger sustained winds. Although lightweight, propeller materials such as aluminum, fiberglass, and carbon fiber still require sustained wind to operate.

Early propellers were "fixed" at a particular pitch. Fixed propellers effectively reduce the range of usable winds for power generation. Light winds will not produce enough revolutions to operate rotors. Alternatively, heavier winds might actually result in very high rpms, which damage wind turbines and towers, or at least accelerate material fatigue.

Propeller engineering has addressed some of the aforementioned issues by creating stronger and lighter weight propellers out of new generation materials and using twenty-first-century production technologies. Propellers can be constructed out of hollow balsa wood frames with well-engineered internal support framing to prevent torque (or twisting) due to wind force impact. Balsa frames are covered with fiberglass to further strengthen the propeller and leading edge caps prevent fiberglass wear or propeller destruction due to impact with flying objects (e.g., birds). With computer wind simulations, it is possible to more efficiently place leading-edge caps to effectively protect propellers from damage while reducing material weight. Computer technology also helps engineers more effectively reduce unwanted wind drag on the propeller system. While drag is critical for the propeller spin, some forms of drag may actually lead to decreased propeller rotation, thus negatively impacting energy production (see Mohamed 2004).

Engineers are increasingly experimenting with the use of carbon fibers as a coating to be combined with fiberglass or used as a sole protective and strengthening layer over balsa wood framing. The logic behind the use of carbon fibers is twofold: (1) carbon fibers are lighter weight than fiberglass and its gel coating; and (2) carbon fibers are more resilient than fiberglass. Recent studies have shown that carbon fiber propellers exhibit some promising results and may reduce replacement and/or operation and maintenance costs for system propellers (see Veers et al. 2003).

Jackson, Zuteck et al. (2005) conducted extensive research, comparing the effectiveness of fiberglass and carbon fiber propeller blades, controlling for blade thickness, length, internal propeller stud support, and chord dimensions. The study found that increased internal supports reduce the required thickness of fiberglass or carbon fiber lamination; reduced thickness translates into reduced weight and material costs in blade manufacturing. Jackson, Zuteck et al. (2005) also studied the impact of various fiberglass and carbon fiber blades, controlling for the aforementioned characteristics, under clean and dirty blade conditions to compare power curves. Blade condition had noticeable impacts on longer blades operating under windy conditions.

Technology has also made it possible for propeller pitch to be adjusted automatically, depending on wind speed, so as to increase system energy production. This means that propellers must be connected to a rotor in a manner that prevents detachment but also permits pitch adjustment while the wind turbine is in operation. Studies have found that propeller fasteners (i.e., bolts) must be designed such that torque specifications are met, but that costs are constrained. Some forms of blade attachment—for example, fiberglass/metal hybrid t-bolts—may be cheaper yet equally effective. Materials analysis is conducted to relate costs to required wind energy system specifications (see Jackson, Zuteck et al. 2005).

Rotor size in relation to wind speed is another important factor in maximizing energy production. The rotor assembly is composed of the propeller blades and motor unit rotated by the impact of wind force on the blades. Rotors on many commercial-grade wind turbines are between approximately 40 and 110 meters in diameter and range in rated power production of between 0.6MW (megawatts) and 4.2MW. The impact of wind on these rotors varies as well, ranging from 312W per one meter squared of wind force to nearly 500W per meter squared of wind force (Jackson, Van Dam et al. 2005). Rotor size is positively related to power production and may reduce the need for larger generators within the rotor assembly (see Jackson, Van Dam et al. 2005). The authors found that the tailoring of wind turbine parameters (e.g., rotor and generator size) to load demands creates greater system efficiency, reduces costs, and increase power generation revenues. Technology is making it increasingly possible for wind turbine peak production periods to match peak load demand through proper turbine placement.

Nevertheless, Griffin (2001) found that rotor sizes are pushing the limits of current wind energy system materials. As blades become larger, some exceeding 60 meters (197 feet) in length, blade materials and blade tip caps must be made thicker, thus increasing blade weight. Heavier blade materials translate into more wear on rotor assemblies and taller and heavier support towers. In order to maintain system efficiency, lighter weight materials must be developed that can withstand the increased demands on wind generation systems as they become larger. As noted in Malcolm and Hansen (2003: 12), an optimum wind tip speed/wind speed ratio is between 7.5 and 8.0; this ratio cannot be maintained using existing materials as rotor radius is increased.

Larger and cheaper energy systems require more than simply identifying next generation materials. New tools and manufacturing processes must be developed to build the next generation of wind energy systems. Thus, the path toward increased use of wind energy systems, particularly large wind energy systems, will require continued commitment to research and development. New-age materials will only come to fruition with increased availability of money needed to support development. Cash resources will either come from existing sales of wind energy systems or from government or privately sponsored research. The transition to large-scale wind energy will require time to develop the next generation of materials, manufacturing processes, and operational design (see Griffin 2001).

At times, new materials and system design are at odds with one another, and a new synthesis on the interaction of design and materials must be completed. In his holistic analysis of wind systems, Ahlstrom (2006) points out that lighter, more flexible materials used in the construction of towers and blades may decrease system efficiency. Flexibility in construction may reduce costs and the impact of wind on equipment failure, but it can also lead to reduced energy generation. There is an interaction effect of blade design and other critical parts of wind turbine systems. Effective designs cannot deal with the individual parts in isolation but must consider the whole requiring the use of computer simulation analysis to isolate the dynamic qualities of wind turbines under different operating conditions and to determine optimum system design and adjustment. Simply put, "A wind turbine is a complex system working a complex environment. It is composed of subsystems working a tightly coupled way" (Diveux et al. 2001: 153).

Beyond the complexities of the individual wind energy system, there is the larger issue of how many wind systems should be deployed at any given generation site. In a study conducted in Scandinavia, Holttinen (2005) found that increasing the number of turbines in a single location does not necessarily improve the ability of wind systems to meet peak load demand. Rather, more turbines in a single location produces greater variation in wind energy production—variation roughly translates into greater uncertainty about the ability of wind generation at a particular site to meet demand. Through a longitudinal study of hourly wind power variations, Holttinen (2005) found that smoothing energy generation, thus reducing variation in expected wind energy production, is best accomplished through greater geographical diversity in wind turbine sitting. Single-site wind energy production methodologies make variation in wind speed, and hence energy production levels, an often variable and unmodeled cost that must be borne by wind energy producers. Conversely, multiple wind energy production sites take advantage of the variability that naturally exists in wind speeds over a large geographical area. Additionally, a smaller number of turbines at any given energy production site reduces the "footprint" of production operation.

Another technological solution to wind speed variation[1] and its impact on wind energy production is energy storage and selective release on demand. Energy storage generally means battery storage. Over-sizing wind turbines to not only meet live-time peak load demand, but to also meet battery storage load demands, is an efficient wind energy design. Large variation in wind speeds, however, will not necessarily be compensated for by over-sizing a wind turbine system or by adding additional turbines. A more effective method of reducing load variations requires an enlarged energy storage capacity (i.e., increased battery storage). In small wind energy systems, Paatero and Lund (2005) found that energy storage reduced power fluctuations by 10 percent, simply by using a small 3kWh (kilowatt hour) storage device.

Wind energy turbines are increasingly located in a variety of different locations, such as coastal waters and mountainous terrain. The advantage of the using these out-of-the-way locations is that they are out of the way. Placed in locations that are not close to homes and businesses, wind turbines are less likely to be opposed by property owners. A major technical difficulty in placing the turbines in these locations, beyond the initial construction issues, relates to the ability of wind energy engineers to correctly site wind turbines. Wind patterns along coastal waters may be difficult to measure. The confluence of wind and ocean currents and water temperatures impact wind patterns. Similar problems face wind energy development in high mountain locations, which may be impacted by wind following mountainous ravines and impacted by air temperature variations.

Technology has made it possible for wind engineers to overcome the obstacles faced by nonconventional energy sites. In Hasager et al. (2005), the authors identify four major remote sensing technologies employed in measuring wind patterns for offshore wind farm siting and development:

1. passive microwave
2. scatterometers
3. radar altimeters
4. synthetic aperature radar (SAR)

In all cases, the systems are not active methods of measuring wind speed. Analog systems usually involved a rotating device and the analysis of revolutions per second to measure wind speed. Passive wind speed measurement evaluates other dynamics related to wind speed without directly measuring the wind itself. Passive microwave, for instance, can be used to measure the impact of wind on a "wind driven ocean surface" from several different measurement points. A data matrix is created to analyze data patterns to determine speed of wind and direction of wind currents (see Piepmeier and Hass 2002). Scatterometers use satellite tracking to measure movement in small waves along ocean surfaces; using known mathematical algorithms for wave movement and wind speed, it is possible to determine wind speeds at different locations. Altimeters measure pressure changes associated with changes in altitude or high or low pressure fronts moving in a particular area. Data measurements on pressure changes in a particular area can be applied to a mathematical algorithm that gauges wind speed as a function of atmospheric pressure changes.

Imaging SAR has frequently been used to measure the topography of various locations on the Earth's surface. Radar signals are directed at the target location and Doppler returns are measured to determine image topography. When used to measure the topography of ocean currents and small wind-waves, some of the radar signal energy becomes "backscattered"—in other words, the signal return measured has been scattered across the surface being studied. Empirical analysis of backscatter is used to measure instantaneous wind speed along the ocean's surface. Hasager et al. (2005) found that imaging SAR was a valuable method of measuring wind speeds over the ocean, where other forms of wind speed measurement are not readily available.

High-altitude mountain locations also face wind direction and strength analysis challenges. Western regions of the United States are characterized by high-altitude mountain areas, many of which have excellent wind resources. Internationally, both Turkey and countries located in the central Asian region have high-altitude mountain regions with significant proven wind energy capacity. As with ocean wind resources, mountain wind resources require high tech approaches to analyzing wind patterns and sustained wind gusts. Several programs have been developed to measure surface temperature and wind patterns controlling for topographical characteristics (see Eidsvik et al. 2004, Eidsvik 2005). Understanding turbulent air flows around mountains is a critical part of understanding wind energy turbine placement in mountainous regions. Turbulence studies in fluid dynamics have demonstrated that air currents and water currents share similar dynamic qualities (see Belcher and Hunt 1998). Modeling of wind patterns for turbine siting is increasingly feasible, which will increase opportunities for remote wind farm locations and increased energy production.

ECONOMIC FEASIBILITY OF WIND POWER SYSTEMS

The economic feasibility of wind power has improved due to technological advancement in wind energy systems and lighter and stronger materials. Economic efficiencies are also improved by increased demand for wind energy systems. Initial costs are spread over more units produced. This assumes, of course, that a large number of devices of a similar type are produced using standard production techniques and tools. However, as wind systems become more prevalent, there will likely be greater variation in local wind conditions and terrain that will impact the efficiencies of wind energy production. To overcome efficiency limitations, wind energy system producers will have to adapt to more varied demands and fewer wind systems of any particular type will have to be produced; thus, production costs for any particular wind energy system will be spread over fewer units produced. In some cases, wind energy system manufacturing will involve original equipment manufacturing production to meet local wind energy site conditions, as would likely be the case for many sea- or mountain-based wind energy systems. Market forces, government incentives, and/or regulations will likely shape wind energy system producers' choices to produce and price levels for items produced as well as the level of demand for specialized one-of-a-kind wind energy systems. As market demand expands, tools and materials will become more numerous as well as multiple fabrication processes, thus potentially reducing costs for a wider demand. At this point, however, "blade mass and costs scale as near-cubic of rotor diameter" (Griffin 2001: ii).

As noted in figure 6.3—reproduced from Griffin's 2001 study of materials costs associated with different sized wind energy systems—the costs for larger wind energy systems scale up quite rapidly because of the increased need for stronger systems to respond to wind impacts. While energy production increases, blade mass and blade costs increase at an even faster rate.

Radius (m)	Rating (kW)	Area* (m²)	Mass (kg) Blade	Root	Fixed	Prod.	Total	S/kW	S/MWh/yr
23.3	750	66.3	1,577	111	$115	$19,100	$19,215	$76.9	$25.1
32.9	1500	132.6	4,292	243	$520	$51,850	$52,370	$104.7	$31.4
38.0	2000	176.8	6,528	336	$970	$79,230	$80,200	$120.3	$34.9
40.8	2300	203.3	8,010	388	$1,320	$97,495	$98,815	$128.9	$36.6
46.6	3000	265.2	11,783	515	$2,350	$144,910	$147,260	$147.3	$40.8
53.8	4000	353.6	17,961	681	$4,405	$224,395	$228,800	$171.6	$46.0
60.2	5000	442.0	24,869	851	$7,180	$316,590	$323,770	$194.3	$50.8

Figure 6.3. Blade Mass and Cost of Rotors

Rotor costs do not increase appreciably in relation to the size of wind energy system.

The mass (in kg) of the blade and root assembly (the portion of the blade which attaches to the turbines in the rotor assembly) increases quite substantially as the power rating (in kW) increases. While power rating increases by approximately five-fold in the chart above, the combined mass of the blade and root increases by fourteen-fold. In other words, system materials tend to increase nonlinearly, as do blade costs as function of power rating. Smaller systems tend to be more efficient than larger systems. Economic feasibility of large-scale wind power systems, therefore, will largely depend on the ability of technical experts to devise cheaper methods and material for constructing high power rating wind energy blades.

The cost of wind energy in an optimum wind scenario of twelve meters per second will produce energy at approximately 5.1¢ per kWh, which is competitive with the costs of conventional power generation. The caveat to this price estimation, however, is that it assumes that the optimum wind is sustained on a twenty-four-hour basis, which is generally not possible. For grid power generation, this poses a smaller problem, since other forms of power generation can take over when wind energy systems are not meeting load demands. Stand-alone wind power systems, however, will require appreciable extra capacity to produce surplus energy to be stored in batteries or other energy storage systems. The surplus stored power can be used to meet load demand during periods when wind energy systems are not meeting load demands. A wind farm requires housing for power storage devices, such as batteries. Servicing storage devices, buildings, and additional wind turbines will require a staff of employees as well as the costs of disposing of worn storage devices, which may contain environmentally harmful substances, such as acid and lead. These requirements will add a substantial cost to the relatively optimistic 5.1¢/kWh. It should be noted that conventional power generation also produces environmental damage and public health problems, which are generally not added into the "true" cost of carbon-based energy.

FEDERAL WIND ENERGY PROGRAMMING:
HIGHLIGHTING WIND POWER AMERICA

The most recent development in federal wind energy policy is Wind Power America (WPA)—a U.S. Department of Energy renewable energy policy initiative that first emerged in 1999. Consistent with other policy initiatives in the 1990s, WPA is designed as a collaborative policy. It does not impose a rigid top-down policy on states, local governments, and businesses. Rather, it promotes cooperation focusing on the varied energy needs throughout the nation. At the national level, the policy goals are to promote state level policy innovations, protect the energy infrastructure of rural America, improve the "green" energy portfolio at the national level, particularly in relation to defense-related activities, and help build partnerships between utilities across local regions and states (Flowers and Dougherty 2002: 1).

The four major goals listed above reflect future-oriented thinking as well as preserving the best aspects of historically important energy goals—namely, production feasibility and broad distribution infrastructures. The energy paradigm under which the United States currently operates was first formed in the 1930s. Private power generators and providers were producing limited amounts of electrical power. Although demand was lower at the time, the price of electricity was quite high. Under Franklin D. Roosevelt (FDR), massive programs to build an energy infrastructure in the United States increased the supply of fairly cheap electrical energy. FDR's policy initiative was top down. Referring back to early chapters, Ted Lowi might label it "distributive" policy. Albeit costly (particularly in terms of workers' lives), the 1930s paradigm was to a significant degree built on hydropower. Dams were built where hydropower was most feasible and energy distributed across the national electrical grid.

The new energy paradigm is expensive and rapidly evolving. There is potential for a centralized approach, by placing large wind farms on vast acreages of open public lands, but it is more likely that energy policy will be tailored to specific state and local needs, in many instances using existing private land resources. Additionally, the expense of renewable/alternative energy systems means that caution and pragmatism must guide infrastructure choices to create efficient renewable/alternative energy plans.

Of historical energy policy importance, the need to promote rural economic development is an important part of WPA. Rural areas are the source of our domestic food supply. Without a ready supply of reasonably priced electricity, agricultural commodity prices will increase dramatically, damaging many sectors of domestic food production and preparation. Rural economic development, however, has broader implications. Rural areas are microcosms of the suburban and rural/urban interface communities throughout the United States. WPA clearly recognizes the need to promote community efficacy in moving into a collective and individual vision for the future. Energy availability and policy programming are important parts of making future social and economic goals a reality.

Providing for the common defense is a basic principle of government embedded in the U.S. Constitution. Perhaps due to the current war on terrorism and the awareness that military force strength is stretched over many national and international commitments, U.S. policymakers have become increasingly aware of the need to have ready access to energy resources in the effort to maintain national autonomy. In the spirit of promoting continued availability of energy resources for purposes of national defense, WPA seeks to promote renewable energy for appropriate defense policy purposes.

Finally, the promotion of utility partnerships are crucial to the development and maintenance of a more unified energy infrastructure, particularly as energy demand continues to rise and supply issues remain a critical issue for utility suppliers and consumers alike. Energy deregulation is intended to help promote energy efficiency and to open the process to a more diverse group of energy suppliers, such as small renewable/alternative energy concerns as well as individual green energy producers seeking to sell green energy through net metering programs (see Wan and Green 1998).

According to Flowers and Dougherty (2002), there are twelve major operating principles undergirding WPA (see figure 6.4). First, WPA works at the "margins" to promote wind energy. Rather than using a "one size fits all" approach, WPA determines the potential policy environment for wind energy and "should avoid investing in markets that are fully commercial and active" (Flowers and Dougherty 2002: 1). Second, WPA helps state and local governments fully "leverage" their existing energy policy resources so as to best promote community vision (Principle 2) and to develop new partnerships at the local, state, and national levels (Principle 3). Not unlike the so-called reinventing government paradigm principles, WPA works to promote strategic vision and the pursuit of "strategic opportunities" (Principle 4) and to fund innovative pilot programs—the foundation of the policy paradigm promoted during the Clinton presidency (Flowers and Doughtery 2002: 2; Osborne 1990; Osborne and Gaebler 1992). In pursuing opportunities to build new local and state initiatives, WPA promotes efforts to replicate successes through best practices policymaking techniques (Principle 6) and to provide needed education and other resources to make collaborative

WPA Activity Matrix

Figure 6.4.   WPA Activity Matrix
Adapted from Flowers and Dougherty (2002: 3).

efforts feasible (Principles 7, 9, 10, 11, and 12), the latter is consistent with the need to promote democratic and inclusive policy initiatives and outcomes (see Pierce and Lovrich 1987). Finally, WPA identifies particularly "challenging . . . markets" (Principle 8) to promote best solutions to the specific energy needs of a particular state or local area (Flowers and Dougherty 2002: 2).

The rubric presented above best summarizes the WPA model for wind energy. The federal program recognizes the tremendous complexity facing the implementation of effective wind energy programs nationwide. It has a clear goal of producing 5 percent of all electrical energy by 2020, but there is no unified roadmap to reach that goal. Unlike the 1930s energy policy paradigm, this twenty-first-century paradigm accurately reflects the goals of a more pragmatic public-private-intergovernmental model of public policy. There is no one best way; limiting participation and narrowing solution options will not produce better outcomes, and active involvement recognizes complexity and the sense that the energy problem is nationwide and the solutions should reflect the diverse nature of the obstacles that are faced.

STATE AND LOCAL POTENTIAL AND EFFORTS

According to *Wind Energy Potential in the United States* (Elliott and Schwartz 1993), the states with the greatest wind energy resources are in the upper midwest, Texas, and the mountain region. North Dakota has the greatest wind energy potential at 1,210 billion kWh. Texas's (which is #2) and Montana's potential capacity stand at 1,190 and 1,020 billion kWh per year, respectively. Wind energy potential, however, does not necessary equate to wind energy development maximizing potential capacity. California, which has the greatest wind energy development, produces a little over 1 percent of its power needs with wind energy.

Top Ten States in Wind Capacity (in billions of kWh per Year)			
North Dakota	1,210	Nebraska	868
Texas	1,190	Wyoming	747
Kansas	1,070	Oklahoma	725
South Dakota	1,030	Minnesota	657
Montana	1,020	Iowa	551

Source: Elliot et al. 1991.

Three challenges facing wind resource development are: establishment of wind farm infrastructure, power transmission, and political/public support. Wind farm infrastructure relates to the costs associated with the transportation and assembly of wind turbines in remote locations. While the quality of wind resources in the mountain states and upper Midwest is quite substantial, wind energy equipment is shipped long distances from coastal ports and industrial centers to remote sites in these states. For each commercial wind generation unit, large turbine blades, rotors, and tower assemblies arrive disassembled and final assembly must be done on-site, requiring human resources and equipment. Mountain and upper midwestern states work with the WPA program to overcome some of the challenges associated with establishing a wind energy infrastructure. With a lower population density and abundant wind resource potential, developments could prove to be very beneficial in the long run.

Power transmission issues will arise as wind power infrastructure impinges on other policy priorities. If wind resources are developed in remote mountain sites, transmission will require the development of power lines crossing wilderness areas, which would pose serious challenges to environmental quality priorities. Rural areas in the western United States frequently abut public lands; wind power systems might pose risks to the plant and animal species living in public spaces.

Political/public support poses one of the most serious challenges to the development of wind power in the United States. In the mountain state region, for instance, Montana and Idaho have passed state laws supporting private wind turbine development by restricting construction near wind turbines that would negatively impact turbine performance. Wind and solar energy laws in many states provide tax and direct cash incentives to individuals and businesses that develop wind energy infrastructure.

Public support becomes more of a challenge when it comes to property values. For instance, the California coastal region north of Los Angeles has some of the best wind resources in the state, yet the scenic beauty of the region makes wind energy development unlikely; real estate development benefits are greater than the power generation benefit. Coastal areas in Washington, Oregon, Michigan, Maine, Massachusetts, and North Carolina face similar public support challenges.

The greatest wind resource development has occurred in California, but generally not along the California coast. California wind energy capacity is approximately 2,096mW installed wind energy systems. Wind energy development in the top wind capacity states is relatively uneven. North Dakota and Montana have developed only a tiny fraction of their wind resources, while Texas, Iowa, and Kansas are progressing more rapidly toward fully developing their wind energy infrastructure.

In 1981, the United States had only 10MW wind energy generation capacity. By 1982, the capacity had increased to 70MW, a 600 percent increase in wind energy generation capacity. Wind energy capacity stands at 6,740MW as of June 24, 2005 (see www.awea.org/projects/, accessed July 7, 2005]).

CHAPTER SUMMARY

Wind energy capacity in the United States is substantial. The technical feasibility of wind energy development is a function of the materials and engineering capable of meeting the various wind and environmental factors. Large-scale wind turbines are limited by current material strengths. Economic feasibility is a function of capital costs and operations and maintenance expenses. As wind energy increasingly explores challenging environments, such as high altitude mountainous and ocean siting, the technical feasibility that allows for wind energy development impacts the costs associated with establishing a wind farm. Political feasibility constraints may impact access to optimum wind power locations for wind power generation development.

NOTE

1. See Elliott and Schwartz (2005). In the United States, the National Renewable Energy Laboratory actively seeks to improve measurement of wind resources and maximize efficiencies of siting and wind generation plant dimensions related to site-specific conditions. Tower height, for instance, is particularly important in understanding wind resources as well as efficient access.

WORKS CITED

Ahlstrom, A. 2006. Influence of Wind Turbine Flexibility on Loads and Power Production. *Wind Energy* 9: 237–49.

Belcher, S. and Hunt, J. 1998. Turbulent Flow over Hills and Waves. *Annual Review of Fluid Mechanics* 38: 507–38.

Diveux, T., Sebastian, P., Bernard, D., and Pulggali, J. 2001. Horizontal Axis Wind Turbine Systems: Optimization Using Genetic Algorithms. *Wind Energy* 4: 151–71.

Eidsvik, K. 2005. A System for Wind Power Estimation in Mountainous Terrain: Prediction of Askervein Hill Data. *Wind Energy* 8: 237–49.

Eidsvik, K., Holstad, A., Lie, I., and Utnes, T. 2004. A Prediction System for Local Wind Variations in Mountainous Terrain. *Boundary-Layer Meteorology* 112(3): 557–86.

Elliott, D. and Schwartz, M. 1993. *Wind Energy Potential in the United States.* www.nrel.gov/wind/wind_potential.html, accessed November 27, 2005.

Elliott, D. and Schwartz, M. 2005. *Development and Validation of High-Resolution State Wind Resource Maps for the United States*, NREL/TP-500-38127. Golden, CO: National Renewable Energy Laboratory.

Elliott, D., Wendell, L., and Gower, G. 1991. *An Assessment of the Available Windy Land Area and Wind Energy Potential in the Contiguous United States*, PNL-7789. Richland, WA: Pacific Northwest Laboratory.

Energy Center of Wisconsin. 2005. *Parts of a Turbine.* www.ecw.org/windpower/cat2a.html, accessed June 30, 2005.

European Space Agency. 2005. *Scatterometer Design.* earth.esa.int/rootcollection/eeo4.10075/scatt_design.html, accessed July 3, 2005.

Flowers, L.T. and Dougherty, P.J. 2002. *Wind Powering America: Goals, Approaches, Perspectives, and Prospects.* NREL/CP-5-32097. Golden, CO: National Renewable Energy Laboratory.

Griffin, Dayton A. 2001. *Wind PACT Turbine Design Scaling Studies Technical Area 1—Composite Blades for 80- to 120-Meter Rotor.* NREL/SR-500-29492. Golden, CO: National Renewable Energy Laboratory.

Hasager, C., Nielsen, M., Astrup, R., Barthelmie, E., Dellwik, N., Jenson, B., Jorgenson, S., Pryor, C., Rathmann, O., and Furevik, B. 2005. Offshore Wind Resource Estimation from Satellite SAR Wind Field Maps. *Wind Energy* 8:403–19.

Holttinen, Hannele. 2005. Hourly Wind Power Variations in the Nordic Countries. *Wind Energy* 8: 173–95.

Jackson, K. J., Van Dam, C. P., and D. Yen-Nakafuji. 2005. Wind Turbine Generator Trends for Site-Specific Tailoring. *Wind Energy* 8: 443–55.

Jackson, K. J., Zuteck, M. D., Van Dam, C. P., Standish, K. J., and Berry, D. 2005. Innovative Design Approaches for Large Wind Turbine Blades. *Wind Energy* 8: 141–71.

Kanpur, Chandra. 2005. On What Principle Does an Altimeter Work? *Times of India* May 28, timesofindia.indiatimes.com/articleshow/1125635.cms, accessed July 3, 2005.

Larsen, T., Madsen, H., and Thomsen, K. 2005. Active Load Reduction Using Individual Pitch, Based on Local Blade Flow Measurements. *Wind Energy* 8: 67–80.

Malcolm, D. and Hansen, A. 2003. *WindPACT Turbine Rotor Design, Specific Rating Study*, NREL/SR-500-34794. Golden, CO: National Renewable Energy Laboratory.

Mohamed, M. 2004. "3D Woven Carbon-Glass Hybrid Wind Turbine Blades." Presentation at Wind Turbine Blade Workshop. Sponsored by Sandia National Laboratories. www.sandia.gov/wind/2004BladeWorkshopPDFs/MansourMohamed.pdf, accessed March 13, 2006.

Osborne, D. 1990. *Laboratories of Democracy.* Boston: Harvard Business School Press.

Osborne, D. and Gaebler, T. 1992. *Reinventing Government: How the Entrepreneurial Spirit is Transforming the Public Sector.* Reading, MA: Addison-Wesley.

Paatero, Jukka and Lund, Peter D. 2005. Effective of Energy Storage on Variations in Wind Power. *Wind Energy* 8: 421–41.

Piepmeier, J. and Hass, J. 2002. *Ultra-Low Power Digital Correlator for Passive Microwave Polarimetry.* Moscow, ID: Center for Advanced Microelectronics and Biomolecular Research. www2.cambr.uidaho.edu/hips/ulp_polarimetry_correlator .pdf, accessed July 3, 2005.

Pierce, J. and Lovrich, N. 1987. *Water Resources, Democracy, and the Technical Information Quandary.* Millswood, NY: Associated Faculty Press.

Pinard, Jean-Paul, Benoit, Robert, and Yu, Wei. 2005. *A West Wind Climate Simulation of the Mountainous Yukon.* Montreal, Canada: Environment Canada. collaboration .cmc.ec.gc.ca/science/rpn/publications/pdf/paperyukon_19_04 _05. pdf, accessed July 4, 2005.

Veers, P., Ashwill, T., Sutherland, H., Laird, D., and Lobitz, D. 2003. Trends in the Design, Manufacture and Evaluation of Wind Turbine Blades. *Wind Energy* 6:245–59.

Wan, Y. and Green, H. 1998. Current Experience with Net Metering Programs. Conference paper presented at *Windpower '98*, Bakersfield, CA, April 27–May 1, 1998.

WEB SITE

www.awea.org/projects/, accessed July 7, 2005.

Chapter Seven

Geothermal Energy

INTRODUCTION

Geothermal power presents amazing potential for global clean energy production (Ghose 2004). As an energy source, it relies on the Earth's heat without altering the Earth's core temperature. The highest quality geothermal energy resources exist domestically in the southwestern United States, but usable "grades" of geothermal resources can be found throughout the nation.

WHAT IS GEOTHERMAL POWER? HOW DOES IT WORK?

The Earth's mantle is composed of superheated iron and other elements. In some locations, this superheated material rises into the Earth's crust and thus closer to deep Earth groundwater supplies in the Earth's crust. Wells are drilled in areas of accessible geothermal activity and the superheated water known as *brine* is extracted (see figure 7.1). A closed system allows the brine resource to transfer its energy to other low boiling point materials. The low boiling point materials in the secondary stage system are used operate high efficiency turbines that, in turn, produce electrical energy. Because it is a closed system, the brine extracted from the geothermal wells is never contaminated by surface use. The brine resource is then reinjected into deep Earth wells in close proximity to the extraction well.

According to the U.S. Department of Energy (DOE), geothermal resources above 200° C are high-grade geothermal resources. Geothermal resource temperatures can be as high as 700° C. High-grade geothermal resources are abundant in the American west. Medium-grade resources (150–200° C) are

Flash Steam Power Plant

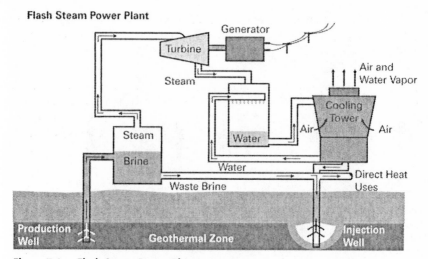

Figure 7.1.   Flash Steam Power Plant
Source: www.geothermal.inel.gov/i/flash.gif, accessed June 10, 2006.

located in the west and southwest, while low-grade resources (100–150° C) are found throughout the United States.

SAFETY, ENVIRONMENTAL DAMAGE, AND EMISSION-RELATED ISSUES: GEOTHERMAL ENERGY

Potentially, geothermal energy resource use can lead to environmental damage and the production of harmful airborne emissions; but problems can be minimized through the use of a closed-loop system, as described in the previous section. A closed-loop system withdraws, uses, and reinjects brine, maintaining it within the confines of piping during the surface process. No brine is pure water. Hydrogen sulfide (H_2S) and other dissolved gases in brine must be trapped in the brine-extraction process to prevent health and environmental damage; this can be accomplished through the use of a closed-loop system. If an open-loop system is used, then brine has contact with surface soils and air, possibly releasing any of several sources of contamination:

- H_2S
- ammonia
- methane
- carbon dioxide
- toxic sludge containing sulfur, vanadium, arsenic, mercury, nickel, etc.

(See: Union of Concerned Scientists 2005 [www.ucsusa.org/clean_energy/renewable_energy_basics/environmental-impacts-of-renewable-energy-technologies.html, accessed November 27, 2005]).

Soil and air contamination are particularly problematic when resource development occurs on public lands. Assuming proper system management, it is possible to maintain a small human technology "footprint" for geothermal resource extraction and use, thus reducing harmful effects on the environment.

TECHNICAL FEASIBILITY OF GEOTHERMAL POWER

Geothermal power can be developed in a variety of ways. In some cases, geothermal energy is used to generate electricity; the technical feasibility of the energy resource is measured in kilowatts (kW_e) or megawatts (MW_e). In other instances, the geothermal resource is used as a source of thermal energy and is measured in kilowatts (kW_t) or megawatts (MW_t). Four widely used types of geothermal technology will be discussed here: heat pumps, direct use, flash steam power, and binary steam power (the two former technologies are thermal, while the two latter technologies are primarily used to generate electricity).

Heat Pumps and Direct Use

Although in use since the 1940s, in the 1970s the technology experienced a surge in consumer interest. Heat pumps have become the most popular geothermal technology in the United States (Lund 2003: 414) and have great potential for reducing thermal heating costs in other nations through large-scale system design (see Bujakowski and Barbacki 2004). Approximately 60 percent of the geothermal energy used in the United States—about 12,500 TJ/yr[1] of energy or 3.47 MW of power—is in the form of heat pump technology. As with other forms of renewable energy discussed thus far, heat pump usage is largely dependent on geography—more specifically, on geological conditions at a specified site and home design considerations (Roman 2004).

Heat pumps are relatively simple devices involving a well, piping, and circulatory pump. For building heating purposes, the completed geothermal well must have ground or water temperatures that are satisfactory to meeting the heating needs of a building. A piping system is lowered into the well and chemicals such as antifreeze are put into the closed piping system (see Zhao 2004). Optimally, there must be no leakage of the heat transfer substance in the piping system into the well's water and surrounding soil resources. A pump at the surface level moves the heated liquid through the closed piping system, and building air is heated by transferring heat to or from the geothermal resource. For instance, in the winter months, cooler air is displaced by the transfer of geothermal warmth to the building's airflow system. Conversely, in the summer months, heat in the building's air can be transferred to the antifreeze or other chemicals in the geothermal piping and pumped into the ground (see Bloomquist 2000).

Heat pumps represent one the largest direct uses of geothermal energy, but there are a variety of other forms of direct use mentioned below. Direct use of geothermal resources worldwide has grown rapidly in recent years and will continue to accelerate in growth (see Fridleifson 2003: 383, 2005). In 1999, direct use of geothermal generated 53 TWh/a of direct use globally (DOE 2003: 216) (see figure 7.2). Direct use resources are often lower quality (lower temperature) resources that can be used for thermal energy purposes such as heating buildings (see Brown 2001) and other heating related purposes.

Fish farming and greenhouse are two increasingly common direct uses of geothermal resource, with resource temperatures 150° C and below (Demĭrbaş et al. 2004). According to Lund (2003), the largest concentration of aquaculture fish farming in the United States is in the Imperial Valley in southern California. Over one million tons of bass, tilapia, and catfish are raised on an annual basis in about a dozen geofarms. Alligators are raised in Imperial Valley commercial farm facilities; the meat and skins harvested are used for food and apparel.

Advanced farming practices are often made possible by the use of geothermal energy. Geothermal energy is used in hothouse farm operations. In shopping for produce at a grocery store, it is likely that one will come across hydroponically raised tomatoes and other fruits and vegetables. Hydroponics requires raising plants in warm water, which is often direct use geothermal water. Geothermal heat resources are also used in dehydration facilities (e.g., dried fruits, vegetables, herbs, and spices). Geothermal energy technology makes it technically feasible to grow and process crops throughout the United States.

While the United States has abundant arable farmland in temperate climates, the technical feasibility of geothermal greenhouses is especially valuable in nations where weather and soil are less conducive to crop production. The geography of Turkey, for instance, is mountainous and the soil and climate has historically made farming a challenge. In a burgeoning society, food resources must keep up with ever-growing demand for long-term sustainability (Kaygusuz and Kaygusuz 2004). Demand for energy has inspired significant energy search processes to maximize geothermal resource potential (see

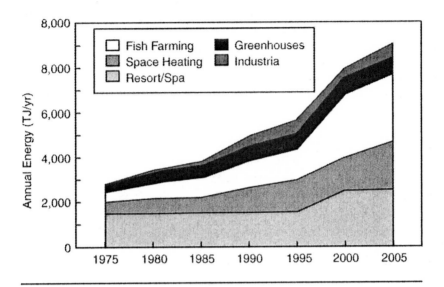

Use	Number of Capacity Installations	Installed Capacity (MWt)	Annual Energy 10⁹ Btu	TJ	Use Factor
Space Heating	975	83	811	855	0.33
District Heating	18	99	592	624	0.20
Aquaculture	53	136	2,819	2,971	0.69
Greenhouses	37	199	1,074	1,132	0.30
Agriculture Drying	3	20	290	305	0.49
Industrial Processing	4	7	73	77	0.35
Resorts/ Spas Pools	219	107	2,369	2,497	0.74
Snow Melting	5	2	16	17	0.27
Subtotal	1,314	573	8,044	8,478	0.47
Geo. Heat Pumps	400,000	4,800	11,385	12,000	0.19
Total		5,373	19,429	20.478	0.25

Figure 7.2.   Geothermal Use, 1975–2005
Source: Geo–Heat Center, Lund and Boyd (2005).

Serpen 2004). In the last twenty years, Turkey has increased its geothermal greenhouse acreage from near zero to over thirty-five hectares (approximately eighty-six English-system acres). Greenhouse acreage must be thought of differently than standard cropland because it is farmed much more intensively and often on a year-round basis. The warmer and more humid greenhouse conditions also allow for optimum crop growth and reduced time to harvest. This means that more food can be produced in a shorter amount of time because of direct use geothermal agriculture technology (Demïrbaş et al. 2004; Ozgener and Gunnur 2004).

Geothermal energy technology is also readily used in the nonfood and leisure/tourism industry. Nonfood industries, such as flower growers, use geothermal aquaculture to produce houseplants and flowers for floral arrangements. Tree seedlings are also developed in geothermal tree farm operations in eight Western states according to EERE (2005 [www.eere.energy.gov/geothermal/directuse.html, accessed November 27, 2005]). Resorts use geothermal energy to create spas and heated swimming pools. In the arid desert southwest, new citizens are often attracted to the hot climate but are likely also to be attracted to warm water in swimming pools and spas, which makes life a bit more enjoyable.

Technologically, it is possible to maintain geothermal aquaculture and agriculture operations as long as geothermal wells are maintained and water temperature managed so that plant life is sustained and not damaged by excessive heat. Well water levels must either be managed through reinjection of surface water or through careful water withdrawal plans. It is important, however, to realize that direct use of geothermal resources remains under study by scientists, determining its availability and the best method of protecting the resource.

Flash Steam Power

High-quality geothermal resources (i.e., very high temperatures and consistent geothermal water availability) emerge from wellheads as steam. The steam can be piped directly into turbine systems. Power take-off from the turbines is controlled so that electrical generators produce usable electrical energy (see Kose 2005: 69)—in the United States, homes and home equipment are wired for 120 volts and 60 Hertz AC current. After steam has been used to operate the turbine, it can be used to operate secondary or even tertiary stage flash steam turbines. When the resource is no longer capable of operating turbine systems, it exits the turbine chamber and is cooled in a condensing system to the point where it becomes a liquid, and is reinjected into the geothermal reservoir through a second well outlet (DOE 2003: 9).

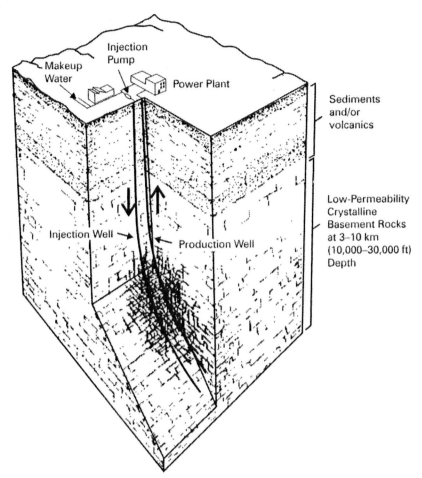

Figure 7.3.   A Geothermal System
Source: Reproduced from Tester et al. (1994: 100).

Reinjected water is cooler than the water than the untapped geothermal resource. The cooler water will impact the average temperature of the geothermal reservoir. Geothermal scientists have been exploring the impact of reinjected water on the geothermal resource in at least two ways. First, scientists
have studied the flow patterns of reinjected water. Second, scientists have
studied the reservoir temperature recovery time of geothermal resources being used for direct use and power generation. In both cases, the studies reveal
interesting facts about the long-term technical feasibility of geothermal resources (see figure 7.3).

In a study conducted in Japanese geothermal reservoirs, scientists injected inert but traceable chemicals into reinjection wells to determine the rate of flow within geothermal reservoirs. The study found that reinjected water can travel quite rapidly once it is reinjected into a reservoir. The cooler reinjected water will mix with hotter geothermal resources and potentially reduce the quality of the resource (Kumagai et al. 2004). Quality reduction, however, has more to do with deep Earth geology surrounding geothermal reservoirs. If the geothermal reservoir is in close proximity to magma veins extending from the Earth's mantle into its crust, then the impact on geothermal resources might be negligible. Nevertheless, if the geological formation surrounding a geothermal resource does not provide a strong heat source, it is likely that heavy use of the geothermal resource will cause fairly rapid decline in temperature.

A secondary issue relates to the availability of adequate geothermal water resources. As water is drawn for direct use or electrical energy production, water resources available in geothermal reservoirs may become depleted. The disposal of treated waste water in the reinjection process has become common in Imperial Valley direct use geothermal projects. In northern California, The Geysers, the only dry steam geothermal reservoir in use in the United States,[2] geothermal electricity generation projects are using treated waste water from nearby cities, such as Santa Rosa, for reinjection. Known as the Southeast Geysers Effluent Recycling Project (SEGEP), waste water is transported from California towns and cities such as Clear Lake, Lower Lake, and Middleton to The Geysers project field and injected into geothermal reservoirs. There is a danger that, despite water treatment, the use of waste water might contaminate geothermal reservoirs (Podger 2003). The contamination concern is heightened by evidence of impurities in the geothermal brine. In 2003, Pryfogle (2005) found that impurities in geothermal water provided the nutrition and warmth necessary for bacterial growth, resulting in a bio-film that formed on equipment (see also DOE 2003: 174).

Despite attempts to monitor geothermal reservoir movement and water levels, The Geysers geothermal electrical generation projects illustrate the long-term problem of heavy use of geothermal resources. Lund (2003: 410) reports that power generation is rapidly declining. Several power generation units in The Geysers project have been shut down due to reduced quality for dry and flash steam power generation. Despite injection projects, power generation for the remaining geothermal units is declining. As of 2003, Lund (2003: 411) reports that 650 MW_e of production capacity have been lost due to declining geothermal resources. The SEGEP injection plan is projected to stabilize geothermal power generation for a short period of time, but then declining resource quality will likely continue.

Binary Systems

The binary geothermal energy system is an alternative method of utilizing geothermal resources. Hot water is drawn up from a geothermal reservoir and is then run through a heat exchanger where the heat from the geothermal resource is transferred to substances with a low boiling point. Time for efficient heat transfer must be carefully determined (see Nowak and Stachel 2005). Typically, organic materials such as pentane or butane are used in binary systems for heat transfer medium. Pentane and butane have low boiling temperatures (36.1° C and −1° C, respectively). Even a low-grade geothermal resource will boil the liquid. The superheated pentane or butane is directed against a turbine generator for electricity production. Following use, the pentane or butane goes through a condenser to be cooled to a liquid form and then returned to the heat exchange unit to be converted back into steam for further electrical power generation (Wicker 2005). The geothermal brine is then reinjected into the reservoir.

As with a flash steam process, the binary system is a closed system. In other words, geothermal water is never exposed to the surface atmosphere. Most importantly, the binary system does not mix geothermal water with the low boiling point chemicals running through the heat exchange unit. Often, binary systems rely on lower quality geothermal resources that do not arrive at the surface in steam form; the geothermal water can be more easily reinjected. In the long run, this means that binary systems will be less likely to reduce the quality of the geothermal reservoir. Improved turbine technologies and the use of high-quality heat transfer, such as isobutane, in the turbine operation process produces higher production efficiencies than have been found in flash steam systems (see reslab.com.au/resfiles/geo/text.html, accessed July 18, 2005).

Technology has made it possible for geophysicists to more accurately identify geothermal potential in regions worldwide, potentially decreasing the need for carbon-based energy sources. For example, at its present rate of growth, India would have to double coal production by 2010 to meet the needs of the population and industry. The resulting carbon monoxide, carbon dioxide, sulfur, nitrogen oxides, and particulate emissions would produce significant air and water pollution. Indian carbon dioxide emissions are projected to reach 700 metric tons annually in 2015. Raghuvanshi et al. (2006) explore the growth of coal-related emissions and potential carbon dioxide sinks—oceans and forests. Geothermal energy in India is a potential method of reducing carbon-based emissions (Ghose 2004).

As with India, Poland is a developing nation experiencing rapid growth in production; between 1999 and 2004, Polish GDP rose by 67 percent. Coal is the primary fuel in electrical energy production—91 percent of fuel used in the generation process, compared with a world average of 37 percent (Rozen and Olejniczak 2005: 2). While coal is readily available, geothermal energy could meet energy demands and reduce carbon-based emissions. Geothermal resources have been more fully explored in Poland since the fall of the Iron Curtain and the introduction of Western geophysical technology. The southern Poland region along the Carpathian Mountain range has high quality geothermal resources. Polish geothermal resources in the region could potentially make the nation an energy exporter in Europe (see Bujakowski and Barbacki 2004).

Technical developments have also led to the standardization of geothermal power plants. Well development (see Garcia-Valladares et al. 2006), well casing (see Thomas 2003),[3] and turbine system corrosion resistance strategies (see Kubiak and Urquiza-Betran 2002), and operations and maintenance have become standardized and globally-based technologies.[4] Turbine systems do not have to be developed on-site; rather, the systems are now produced using standardized technology and are self-contained "turn key" units capable of being easily transported to the geothermal site for ready use. Technical support is widely available. Energy systems are capable of being monitored globally via satellite and Internet technology.

Technical feasibility is, however, ultimately a function of the continued quality and quantity of geothermal resources. It is possible that overuse of geothermal resources will reduce the quality of reservoirs either permanently or for significant periods of time. In 2003, Ladislaus Rybach published a study related to geothermal sustainability. Rybach found that heat pump impact on geothermal resources was reversible, but required significant recovery time. Rybach found that the quality of the geothermal resource declines over a thirty-year period. Recovery of the resource occurs over a sixty-year period. Rybach calls for continued monitoring and study of large-scale power production. The study illuminates the fact that geothermal technical feasibility must take into consideration the resource quality; feasibility of the resource development must be done with consideration to *sustainability*, thus avoiding exploitation and reduced availability for future generations.

ECONOMICS OF GEOTHERMAL POWER

Economic calculations for geothermal power vary substantially. Estimated costs are a function of the quality of the geothermal resource (reservoir flow rate and pressure, geological fractures), location, demand level (directly related to market price), type of energy use (i.e., electrical or thermal uses), and project development (see Tester et al. 1994: 119). High-quality resources whose development is less costly will reflect the cost savings in kW_e/h or kW_t/h cost structures. Conversely, resources that are more difficult to develop and access will result in higher costs.

One approach to dealing with resource quality, particularly resource quality variation in a single reservoir region, is known as cycling. Cycling limits continual use of any single geothermal resource. The method relies on the use of multiple technologies and power generation scenarios to maximize the economic benefits and protect geothermal resources for future use (see Bloomfield and Mines 2002).

The Renewable Energy Policy Project (REPP)[5] provides some data on the cost of geothermal energy development. The REPP study was conducted in 1999, so costs have been adjusted for 2005 constant dollars, with the assumption that costs have generally remained the same for various geothermal plant sizes. Small energy production of high-quality geothermal resources is cost effect for the residential user. For plants smaller than 5MW, total capital costs are estimated at between $1,858 and $3,484/kW depending on the quality of the geothermal resource.[6] Large plant capital costs are much lower on a per kW basis—between $1,336 and $2,255/kW. In other words, a 30mW plant would cost $40–76 million in initial capital costs. Operations and maintenance costs for geothermal energy production range between 0.5 cents/kWh and 1.5 cents/kWh. Taking into account capital and operations and maintenance costs, therefore, the cost of energy for geothermal range between $1.75 and .08 cents per kWh

Geothermal energy used by individual homeowners is often in the form of heat pump technology. Heat pump technology uses existing heating and cooling ducts, reducing installation costs. According to Canadian analysis, the use of geothermal heat pump technology can reduce the capital costs associated with home energy use by as much as 50 percent (Office of Energy Efficiency 2006). Additionally, energy production costs can be as little as one-third the costs associated with traditional fossil energy-based heating and cooling systems.

It is important to realize that many cost estimates associated with geothermal energy are best case scenarios (see Gelegenis 2004). An example from a World Bank-funded project in Lithuania sheds some light on the challenges that might be faced by future geothermal energy development. The impetus for the project was twofold. First, in the early 1990s Lithuania—recently independent following a near fifty-year occupation by the Soviet Union—began to explore the development of clean renewable energy sources. As was the case with many former Soviet republics, Lithuania was forced to deal with many environmental disasters brought about by heavy industrial development and the misuse of natural resources. Second, the World Bank and other globally-focused organizations saw an opportunity to help newly emerging Eastern European democracies rebuild their societies and clean up their environments.

The ensuing geothermal demonstration project in Klaipeda was a collaborative effort involving the government of the Republic of Lithuania, Enterprise Geoterma, Global Environment Facility, the Danish Environmental Protection Agency, and the World Bank. The project was intended to provide district heating and energy services, producing both thermal and electrical energy for local consumers. Klaipeda hit two major snags in development, one of which continues to plague the geothermal system. First, the temperature of the geothermal resource was slightly lower than had been anticipated. The geothermal well is 1,300 meters in depth, but the water resource temperature is about 38–39° C rather than 40–42° C (see World Bank 2003: 7). Additionally, the water flow rate for the resource well was lower than expected, which required the drilling of a secondary well in hopes of accessing a better quality geothermal resource. At reinjection, the geothermal brine temperature was approximately 17° C, which may pose a long-term resource quality problem, as noted in the technical section of this chapter. The initial reinjection well was incapable of handling the water flow and second reinjection well was drilled. The resource temperature and well-quality issues were dealt with effectively; however, the cost of the project increased.

A second problem relates to mineral concentrations in geothermal brines. Brine can contain high concentrations of gypsum or hydrated calcium sulfate—a substance commonly found in wallboard and in alabaster decorative carvings. Gypsum has a tendency to clog the geothermal piping system and increase breakdown rates for geothermal energy development.

The future of the Lithuanian project is uncertain. With the increased costs of natural gas, geothermal is a more viable option for district energy provision. A zero-emission energy source, geothermal power will also alleviate air, water, and soil pollution in Lithuania and reduce health risks associated with fossil energy use. Nevertheless, economic analysis is driven by two very important assumptions: sustainability of the geothermal resource and low system breakdown rates. At the moment, neither assumption is playing out as hoped. The geothermal resource is weaker than originally anticipated; the World Bank (2003) report indicates that this goal will be "marginally satisfactory[ily]" met. Gypsum clogging has not been overcome—the World Bank report concludes that unless this problem is solved, the system will not be more economic than the current natural-gas-based system.

Klaipeda offers many important lessons for the development of geothermal energy as a sustainable resource. The World Bank project analysis demonstrates that unified support from local, national, and international entities was not present which may have made implementation of the geothermal demonstration project more challenging. The Lithuanian case demonstrates that, on paper, geothermal might be technically, economically, and politically feasible, but commitment must be strong and consistent, particularly when facing changing and challenging circumstances. Technical feasibility aside, without unified support and sustainable economic and political support, a resource as widely available as geothermal energy will not be accessed and used to full potential. Worse still, resource exploitation may result in destruction of the resource.

FEDERAL GEOTHERMAL ENERGY PROGRAMMING

Federal involvement in geothermal energy is focused on three major areas: (1) direct and externally funded research and development programs; (2) monitoring geothermal energy demand, supply, and cost; and (3) intergovernmental and public-private partnerships. Federal programs are designed to promote geothermal energy development through a well-regulated and efficient market environment (see also Bustamante 2000). In the United States, the federal government monitors the use or misuse of geothermal resources so as to maintain resource quality and availability. Through the three-pronged approach listed above, the federal government seeks to minimize its role in the development of geothermal energy resources, but maintain a role in promoting the growth of geothermal energy. Energy policy in the area of geothermal resource development is often intended to shape public-private institutional arrangements, leading to greater energy independence and the promotion of public health through clean energy usage.

Since 1971, the DOE and related organizations have actively pursued domestic geothermal R & D projects (DOE 2001: 1–3). The new millennium has witnessed many developments, not the least of which has been growing interest in the use of geothermal energy for commercial as well residential purposes. The goals of federal government in terms of geothermal energy reflect the noteworthy change. In *Federal Geothermal Programming Update for Fiscal Year 2000*, DOE indicates the following mission and goals for geothermal energy development in the new millennium:

Mission: To work in partnership with U.S. industry to establish geothermal energy as an economically competitive contributor in the U.S. energy supply.

Goal 1: Double the number of states with geothermal electric power facilities to eight by 2006.

Goal 2: Reduce the levelized cost of generating geothermal power to three–five cents per kWh by 2007.

Goal 3: Supply the electrical power or heat energy needs of 7 million homes and businesses in the United States by 2010 as well. Private-sector inputs to DOE's planning process are critical to logical, balanced strategy for the Geothermal Program (DOE 2001: 1–3).

The DOE Geothermal Program

- Conversion Technology
- Drilling Technology
- Energy Systems
- Exploration Technology
- Geo-Powering the West
- Geosciences Projects
- Reservoir Technology

Policy goals are often best measured by resource allocations. Figure 7.4 is a four-year analysis of federal geothermal grant giving and cost sharing ordered according to federal grant priority. It is my contention that prioritizing

Year	Grant Focus	Government Grant Amount	Cost Sharing Amount	Ratio Gov't/Cost Sharing
2000	Drilling	$4,628	$1,515.73	3.05
2000	Exploration	$4,310	$88.76	48.56
2000	Conversion Technology	$3,969	$1,368.54	2.90
2000	Reservoir	$2,887	$32.58	88.59
		$15,794	$3005.61	5.25
2001	Drilling	$6,373	$2,612.45	2.44
2001	Geosciences Projects	$5,931	$1,242.36	4.77
2001	Energy Systems	$4,950	$1,039.30	4.76
2001	Exploration	$3,537	$532.75	6.64
2001	Geo-powering the West	$2,183	$00	.00
		$19,974	$5,426.86	3.68
2002	Geosciences Projects	$7,459	$515.05	14.48
2002	Drilling	$6,140	$2,294.62	2.68
2002	Energy Systems	$3,265	$1,173.12	2.78
2002	Exploration	$1,326	$333.33	3.98
		$18,190	$4,316.12	4.21
2003	Drilling	$5,957	$1,236.59	4.82
2003	Energy Systems	$5,314	$1,710.83	3.11
2003	Reservoir	$4,479	$6,738.17	.66
2003	Exploration	$4,349	$504.73	8.62
2003	Geo-powering the West	$2,564	$583.07	4.40
		$22,663	$10,773.39	2.10

Figure 7.4. Federal Grants for Geothermal Projects, 2000–2003

money speaks volumes about the nature of energy development. The figure illustrates a tremendous focus on drilling technology. Drilling studies have focused on the use of acoustic technology to guide drilling toward optimum locations within a geothermal reservoir. Studies also focused on drill bit strength when drilling through ultra-hard rock formations and the impact of mud and tailings on drilling systems. After a well is drilled, and even during the process of drilling, cement casing is often used to maintain the integrity of a geothermal well. Given the mineral concentrations often found in geothermal brine, drilling studies also focused on the ability of various types of cement to withstand high temperatures and mineral deposits. Except for 2002, drilling technology studies were the top recipients of grant dollars.

Reservoir-focused research, although largely unfunded in 2001 and 2002, witnessed the greatest percentage increase in adjusted dollar grants—an approximately 55 percent increase in federal grant funds were awarded to reservoir studies. Reservoir studies focused on issues such as the detection, location, and description of geothermal reservoirs, critically important information for assessing project development and sustainability. While drilling studies were heavily funded by federal grant dollars, reservoir studies were increasingly funded by cost sharing with state or local governments or with private industry. Increased emphasis on cost sharing is perhaps recognition of mutual or shared interest in research outcomes. Obviously, maintaining reservoir quality ensures maximum lifespan of a geothermal energy project. Exploration studies also experienced increased cost sharing as geothermal projects have become of greater interest to private industries as an energy source. In 2002, government grants as a proportion of funding increased but have subsequently fallen to a four-year low ratio of $2.10 in government funding for every $1 in cost-sharing monies. In fiscal year 2000–2003, the government was seeking to provide a credible commitment to an energy source and private industry becoming more focused on using that energy source.

Federal geothermal programs extend beyond research and development grants to universities, national research labs, and private enterprises. Under 26 USC §48, Renewable Energy and Production Tax Credit (enacted in 1978), corporations that "invest in or utilize solar or geothermal energy property in the United States" are eligible for a 10 percent tax credit (maximum $25,000 per year) and an additional 25 percent tax credit of the remaining total tax (Database of State Incentives for Renewable Energy 2005 [www.dsireusa .org/library/includes/incentivesearch.cfm?Incentive_Code=US02F&Search= Technology&techno=geothermalelectric&currentpageid=2, accessed July 31, 2005]). Tax expenditures for government, credits tied to specific policy goals often create incentives for private sector developments. Industries begin to develop as demand for alternative energy technologies begin to grow and, at least up to a point, average production cost curves for alternative energy manufacturers and installers are impacted by increased unit production and sales.

STATE AND LOCAL EFFORTS

State and local research and development efforts are often financed by private development money or federal research grant money as discussed in the previous section. Some states, such as Maine, offer a matching fund program, essentially matching funds for renewable resource projects, such as geothermal projects, related to specific public institutions (e.g., the University of Maine, Maine Maritime Academy, and community renewable energy demonstration projects). Most states use a system of tax credits as incentives for private development of geothermal and other renewable energy resource projects. Renewable portfolio standards at the state level create a market for "green energy"—a market that is increasingly global (see Goken et al. 2004: 442)—by requiring the use of green energy (zero- or low- emission-produced energy). Tax incentives for geothermal thermal projects, such as heat pumps, are quite common. Additionally, public utility commissions and power companies offer green energy generation incentives for renewable energy providers such as commercial and small production geothermal energy generators in the form of property tax credits or net metering programs that pay green energy producers for the zero or low emissions credit for energy produced. Green tag programs and free market trading possibilities are additional incentives which exist at the state and local level and are used by utilities and businesses to meet green energy requirements (see Victor 1998).

Geothermal energy is a widely available energy resource in the United States, although the highest quality deposits are in the western and southwestern states such as California, Oregon, Washington, Arizona, Nevada, and Utah. Unlike wind and solar energy, geothermal is largely invisible to the public, as the resource is subterranean and the footprint for energy production equipment has become compact.

Since the passage of State Assembly Bill 1905 in 1981, which promoted the development of geothermal resources, California has developed geothermal direct use and electrical generation facilities scattered in cities and towns throughout the state (www.energy.ca.gov/geothermal/, accessed July 31, 2005). Test wells are being explored in northern Arizona and in eastern Nevada for future geothermal resource development (EERE 2005 [www.eere.energy.gov/geothermal/solicitations_awards.html, accessed July 31, 2005]). A cursory ten-year review of major news sources produced not a single instance where geothermal power development was protested. However, the state of Hawaii published a public opinion study indicating some public concern about the potentially negative impact of geothermal brine and aerial emissions on local community health and indigenous theological traditions (www.state.hi.us/dbedt/ert/geo_hi.html# anchor351481, accessed July 31, 2005).

The Hawaii public opinion survey does offer some points for consideration. Solar and wind energy are common resources shared by all members of a community, and it is difficult to conceive of methods by which individuals would be denied access to sun or wind. In the case of geothermal energy, the resource could easily be thought of as a common resource. Geothermal energy traces its origins to the planet's core temperature and shifting deep Earth geological processes. It is a source of energy that is needed by all people and by flora and fauna. Although clearly a resource accessed and developed privately, the existence of the resource has common benefit. In that sense, one could conceive of it as a marketable public good—but a public good nonetheless. A discussion of sustainable communities and energy would be incomplete without at least highlighting this issue, made evident to some degree from the results of the Hawaii public opinion study. As geothermal energy becomes more important to sustainability, the marketable public good dimension of geothermal energy will likely take on greater meaning.

CHAPTER SUMMARY

Geothermal energy sources in the United States are ubiquitous, but are of varying levels of quality. Of minimal impact to resources, geothermal heat pumps are of growing interest to homeowners. Heat pumps can significantly reduce the costs of heating a home. Using higher-quality geothermal resources, thermal systems are capable of producing large quantities of energy for use in operation of private business enterprises, homes, and office space. Geothermal electric systems are capable of generating power to be sold by power companies as green energy; additionally, independent green energy power producers (e.g., geothermal energy power producers) can receive green tags, certifying their green energy production capacity. Green tags are sold on the open market to commercial power producers that can claim the green energy as part of their energy portfolio to meet green energy production requirements.

The future of geothermal energy is a function of the ability of technical experts to locate high-quality geothermal resources and effectively utilize them without depleting the resource. The economics of geothermal energy production are generally straightforward. Geothermal energy costs are competitive with other energy sources. However, without proper understanding of a geothermal site, it is possible for geothermal to become cost ineffective. The Lithuanian case study provides ample evidence of the need to form strong links between the technical, economic, and political dimensions of project development. Understanding and managing geothermal energy are critical aspects of developing a sustainable resource.

NOTES

1. TJ stands for terajoules.

2. Dry steam resources are very rare. The resource is accessed through wells between 7,000 and 10,000 feet deep. The high temperature resource is brought to the surface as steam and is used to operate turbine generators. In a hot water or flash steam system, the resource is brought to the surface as hot water and then through lowering pressure of resource, the water flashes into steam used to operate turbine generators for electricity.

3. Thomas (2003) discusses the use of titanium as a well-casing material due to its ability to resist the corrosive capacity of certain geothermal brines.

4. Prasad et al. (1999) developed a "neural network" model of plant operation that focused on how human operators at a plant interact and make decisions regarding plant operations.

5. REPP is a nonprofit organization that is funded by federal and private grants. One major federal funding agency is the National Renewable Energy Laboratory.

6. The cost effectiveness for small plants is largely a function of keeping down the costs of well development. In remote locations, geothermal plant operation must be scrutinized before purchase to ensure that it has a low breakdown rate and is capable of operating under location conditions. Small geothermal plants are an important method of providing distributed power to remote locations, particularly in developing nations (see Vimmerstedt 1999).

WORKS CITED

Bloomfield, K. and Mines, G. 2002. Predicting Future Performance from Reservoir Management Cycling. *Transactions* 26: 22–25.

Bloomquist, R. 2000. Geothermal Heat Pumps: Five Plus Decades of Experience in the United States. *Proceedings World Geothermal Congress*, pp. 3373–78.

Brown, B. 2001. Klamath Falls Geothermal District Heating System Flow and Energy Metering. *GHC Bulletin* June: 10–11.

Bujakowski, W. and Barbacki, A. 2004. Potential for Geothermal Development in Southern Poland. *Geothermics* 33: 383–95.

Bustamante, C. 2000. PNOC Geothermal Projects: A Holistic Approach to Environmental Management. *Proceedings World Geothermal Congress*, pp. 539–44.

Database of State Incentives for Renewable Energy. 2005. *Business Energy Tax Credit*. http://www.dsireusa.org/library/includes/incentivesearch.cfm?Incentive _Code=US02F&Search=Technology&techno=geothermalelectric&currentpage id=2, accessed July 31, 2005.

Demïrbaş, Ayhan, Şahin-Demïrbaş, Ayşe, and Demïrbaş, A. 2004. Turkey's Natural Gas, Hydropower, and Geothermal Energy Policies. *Energy Sources* 26: 247–48.

Energy Efficiency and Renewable Energy (EERE). 2005. *Direct Use of Geothermal Energy*. www.eere.energy.gov/geothermal/directuse.html, accessed November 27, 2005.

Fridleifson, I. 2003. Status of Geothermal Energy amongst the World's Energy Sources. *Geothermics* 32: 379–88.

Fridliefson, I. 2005. Geothermal Energy amongst the World's Energy Sources. *Proceedings: World Geothermal Congress* April 24–29, pp. 1–5.

Garcia-Valladares, O., Sanchez-Upton, P., and Santoyo, E. 2006. Numerical Modeling of Flow Processes Inside Geothermal Wells: An Approach for Predicting Production Characteristics with Uncertainties. *Energy Conversion & Management* 47(11/12): 1621–43.

Gelegenis, J. 2004. Rapid Estimation of Geothermal Coverage by District-Heating Systems. *Applied Energy* 80: 401–26.

Ghose, M. 2004. Environmentally Sustainable Supplies of Energy with Specific Reference to Geothermal Energy. *Energy Sources* 26: 531–39.

Goken, G., Ozturk, H., and Hepbasli, A. 2004. Geothermal Fields Suitable for Power Generation. *Energy Sources* 26: 411–51.

Kaygusuz, K. and Kaygusuz, A. 2004. Geothermal Energy in Turkey: The Sustainable Future. *Renewable and Sustainable Energy Reviews* 8: 545–63.

Kose, R. 2005. Research on the Generation of Electricity from Greenhouse Resources in Simav Region, Turkey. *Renewable Energy* 30: 67–79.

Kubiak, J. and Urquiza-Betran, G. 2002. Simulation of the Effect of Scale Deposition on a Geothermal Turbine. *Geothermics* 31(5): 545–62.

Kumagai, N., Tanaka, T., and Kitao, K. 2004. Characterization of Geothermal Fluid Flows at Sumikawa Geothermal Area, Japan, Using Two Types of Tracers and an Improved Multi-Path Model. *Geothermics* 33: 257–75.

Lund, John W. 2003. The USA Geothermal Country Update. *Geothermics* 32: 409–18.

Lund, J., Bloomquist, G., and Renner, J. 2005. The United States of America Country Update. World Geothermal Congress, International Geothermal Association.

Lund, John W. and Boyd, T. 2000. Geothermal Direct-Use in the United States. *Geo-Heat Center Bulletin* (Klamath Falls, ID), Geo-Heat Center 21(1): 1–16.

Nowak, W. and Stachel, A. 2005. Assessment of Operation of an Underground Closed-Loop Geothermal Heat Exchanger. *Journal of Engineering Physics and Thermophysics* 78(1): 136–43.

Office of Energy Efficiency. 2006. *Heating Energy Cost Comparison: Heat Pump and Electric Heating Systems*. Office of Energy Efficiency, Natural Resources, Canada. http://oee.nrcan.gc.ca/publications/infosource/pub/home/heating-heat-pump/ heccomparison.cfm?attr=4, accessed March 19, 2006.

Ozgener, Onder and Gunnur, Kocer. 2004. Geothermal Heating Applications. *Energy Sources* 26: 353–60.

Podger, P. 2003. Geysers Wastewater Project Begins an Uphill Run Today. *San Francisco Chronicle*, December 3, A23.

Prasad, G., Swidenbank, E., and Hogg, B. 1999. A Novel Performance Monitoring Strategy for Economical Thermal Power Plant Operation. *IEEE Transactions on Energy Conversion* 14(3): 802–9.

Pryfogle, P. 2005. Monitoring Biological Activity at Geothermal Power Plants, INL/EXT-05-00803. Idaho Falls, ID: Idaho National Laboratory.

Raghuvanshi, S., Chandra, A., and Raghav, A. 2006. Carbon Dioxide Emissions from Coal-Based Power Generation in India. *Energy Conversion & Management* 47(4): 427–41.

Roman, H. 2004. The Earth Can Heat Our Homes. *Power & Energy*. www.techdirections.com, accessed March 19, 2006.

Rozen, K. and Olejniczak, P. 2005. Poland's Coal-Based Power Generation: Will Pressure Create Diamonds? *World Power 2005*, pp. 1–5.

Rybach, L. 2003. Geothermal Energy: Sustainability and the Environment. *Geothermics* 32: 463–70.

Serpen, U. 2004. Hydrogeological Investigations on Balçova Geothermal System in Turkey. *Geothermics* 33: 309–55.

Tester, J., Herzog, H., Chen, Z., Potter, R., and Frank, M. 1994. Prospects for Universal Geothermal Energy from Heat Mining. *Science & Global Security* 5: 99–121.

Thomas, R. 2003. Titanium in the Geothermal Industry. *Geothermics* 32(4–6): 679–87.

Union of Concerned Scientists. 2005. *Environmental Impacts of Renewable Energy Technologies*. www.ucsusa.org/clean_energy/renewable_energy_basics/environmental-impacts-of-renewable-energy-technologies.html, accessed November 27, 2005.

U.S. Department of Energy (DOE). 2001. *Federal Geothermal Research Program Update Fiscal Year 2000*. Idaho Falls, ID: U.S. Department of Energy, Idaho Operations Office.

U.S. Department of Energy (DOE). 2002. *Federal Geothermal Research Program Update Fiscal Year 2001*. Washington, D.C.: U.S. Department of Energy, Energy Efficiency and Renewable Energy, Geothermal Technologies Program.

U.S. Department of Energy (DOE). 2003. *Federal Geothermal Research Program Update*, DOE/NE-ID-11147. Washington, D.C.: U.S. Department of Energy.

Victor, D. 1998. Green Markets. *Ecology* 79(6): 2210–11.

Vimmerstedt, L. 1999. Opportunities for Small Geothermal Power Projects. *GHC Bulletin*, June, pp. 27–29.

Wicker, K. 2005. Geothermal: Hotter than Ever. *Power* 149(1): 40–44.

World Bank 2003. *Implementation Completion Report (CPL-40130) on a Loan in the Amount of US$ 5.90 Million Equivalent and a Grant from the Global Environment Facility in the Amount of SDR 4.60 Million (US$ 6.90 Million Equivalent) to the Republic of Lithuania for a Klaipeda Geothermal Demonstration Project*. Washington, D.C.: World Bank.

World Energy Council. 2005. *Geothermal Energy*. London: World Energy Council, www.worldenergy.org/wec-geis/publications/reports/ser/geo/geo.asp, accessed July 17, 2005.

Zhao, L. 2004. Experimental Evaluation of a Non-Azeotropic Working Fluid for Geothermal Heat Pump System. *Energy Conversion and Management* 45: 1369–78.

WEB SITES

reslab.com.au/resfiles/geo/text.html, accessed July 18, 2005.

www.energy.ca.gov/geothermal, accessed July 31, 2005.

www.geothermal.inel.gov/i/flash.gif, accessed June 10, 2006.

www.state.hi.us/dbedt/ert/geo_hi.html# anchor351481, accessed July 31, 2005.

Chapter Eight

New Century Fuels and Their Uses

INTRODUCTION

Energy storage is an important part of the current and future energy infrastructure. There are two familiar forms in which energy can be stored: (1) battery storage and (2) fuel storage. New forms of energy storage, such as the use of ultra capacitors (see Ashtiani et al. 2006), will become readily available. At the moment, however, batteries and fuels are the familiar elements in the storage paradigm. Advances in battery technology have been tremendous, because of research on material, storage, and discharge capacities (see Jacoby 2006). Programs such as the National Renewable Energy Laboratory's Electric and Fuel Cell Vehicles Program and the Batteries for Advanced Transportation Technologies (BATT) Program operated cooperatively by Lawrence Berkeley National Laboratories and the U.S. Department of Energy (DOE) have pursued next-generation battery research and development. Battery research, however, is only one dimension of the energy future.

A new fuel paradigm is a critical part of the energy future. A useful fuel is one that can be effectively stored and, in a stored setting, is accessible for use on demand. Coal is a very large portion of our fuel paradigm, prominently used in the generation of electricity. Heavy reliance on petroleum for transportation fuel is well known and understood. Price factors, supply forecasts, and concerns with fuel emissions have moved the energy paradigm toward increased diversity in fuel type and source. Natural gas, for instance, will become more prominent, as will alternative fuel development. Knowledge of the status of the fuel paradigm and potential is an important aspect of understanding energy sustainability in communities of the future.

FUEL AS A CONCEPT

Fuel is not an easy concept to define. Technical issues aside, a fuel only becomes an accepted part of human existence when it is economically and politically/socially acceptable. If one lived the Robinson Crusoe existence, then it is likely that fuel would be defined a bit differently: it would be a readily available energy source of reasonable cost. Society's multiplicity of preference sets means that fuel as concept is a much more complicated issue and the words "good," "bad," and "expensive" become modifiers to the word "fuel." Understanding fuel as a concept requires consideration of all three dimensions. As the chapter will spend considerable time discussing fuels, a brief overview of fuel as a concept will be presented at this point.

From a technical standpoint, a fuel is neither desirable nor undesirable. Rather, fuel is *a form of stored energy that may be expended (most commonly through combustion) to release energy*. The ability to access and harness a fuel is the critical first step in making it a useful source of energy for human society. When access to the energy becomes technically feasible, elements and chemical compounds may be designated as a fuel.

In daily existence, most individuals do not think very often about the near-miraculous nature of fuel as stored energy. Yet, accessible fuel sources have been central to human society since prehistoric times and the discovery of fire. Many traditional sources of fuel are still widely used to heat homes and offices and to generate electricity. Wood is probably the earliest-known forms of fuel in human society.[1] Another traditionally used form of fuel still in use today is crude petroleum found in the form of a thick tar called bitumen. As mentioned previously, crude petroleum in the form of tar oozing from the sands was used to caulk the hulls of ships and to serve as energy for cooking.

Over the last two hundred years, technological developments have made energy sources more accessible and methods of production were streamlined. Economically, it became possible to extract, process, and store large quantities of fuel at a reasonable price. Fuel sources became more accessible to consumers. If it were not for these conditions being met, much of modern society would not exist. Revolutionary developments in the petroleum industry and the development of large-scale coal mining in the United States and in other developed nations were critical preconditions to the raising of large-scale, sophisticated militaries, the construction of cities and factories, and the provision of electrical energy to millions of citizens. In fact, reasonably priced stored energy in the form of fuel is often assumed to be an essential good in the modern society (Podobnik 2006).

As discussed previously, however, many things began to change in the 1970s. First, the oil crises demonstrated that reasonably priced accessible fuel sources are not "a given"; as with nearly every other good or service in modern society, fuels are a commodity with price driven by supply and demand. For various reasons, the price of petroleum declined during the 1980s and 1990s, which led many consumers and policymakers to forget about the nature of fuel, mainly crude petroleum, as a commodity. As world demand has grown, partly because of rapid economic development in Asian nations such as China and India, the prices of fuel are increasing. The economics of the fuel sources used for well over a century have changed rapidly.

Second, the 1960s and 1970s brought increased awareness of the pollution produced by the use of traditional fuels such as wood, petroleum, and coal. In all three cases, the use of these energy sources produces quite harmful effects to the biosphere. Critics of the fossil fuel paradigm argue that the true costs of using traditional fuels outweighs the economic and social benefits produced.

Alternative fuels as sources of stored energy are attempts to respond to both issues simultaneously. First, alternative fuel source development will expand the supply of energy available for sale in the marketplace. Basic economics dictates that increased supply with demand held constant will result in decreased prices for fuels. Should demand increase, which will most likely happen, increased supplies of energy might either stabilize prices and price growth or perhaps reduce fuel prices over time. Second, alternative fuels are often researched and produced with the idea of reducing the harmful effects of using conventional fuel sources. As demand and use of conventional fossil fuels expands, so, too, will the quantity of potentially harmful emissions. Alternative fuels, however, offer a range of solutions to the emissions dilemma and each must be considered when making choices regarding an optimum fuel source for the sustainable community of the future.

NATURAL GAS

Natural gas is not normally considered an alternative fuel, because often the word "alternative" is misunderstood to be "renewable" or "green." Natural gas is a hydrocarbon composed primarily of methane and propane. It was used in Britain in the late eighteenth century for commercial and residential purposes. The U.S. natural gas industry began to take form in the early nineteenth century with well digging and gas pipeline developments. In 1859, Edwin Drake, the individual who dug the first oil well in the United States, piped natural gas from the wellhead into a local community for use. As a source of illumination, natural gas was superceded by the advent of electricity. With technological developments in efficient pipeline transportation (e.g., advanced welding techniques, which made pipelines less leak prone), natural gas became an important part of the thermal energy market and was used for residential and commercial purposes. Following the passage of the Natural Gas Act of 1938, rules governing the natural gas industry were promulgated and enforced by the Federal Energy Regulatory Commission.

Natural Gas Composition	
Methane	70–90 percent
Propane	0–20 percent
Carbon Dioxide	0–8 percent
Oxygen	0–0.2 percent
Nitrogen	0–5 percent
Hydrogen sulfide	0–5 percent
Rare gases	trace

The natural gas industry was deregulated in the 1980s and 1990s to encourage further development of natural gas resources. Natural gas's relatively low level of emissions and price were major selling points. Domestic supply has been curtailed by high demand on the existing proven reserves and reduced access to natural gas exploration on public lands. Nevertheless, demand for natural gas is on the rise. Excess demand has led to increased importation of natural gas. For transportation purposes, the gas is cooled and stored in a liquefied form—liquefied natural gas (LNG).

Currently, only six major U.S. ports are capable of handling LNG offloading from transport ships, which reduces supply accessibility.[2] One major LNG port facility areas is along the Texas and Louisiana Gulf Coast region—an area recently hit by Hurricane Katrina. President George W. Bush considers natural gas a critical part of the nation's fuel paradigm and has proposed the expansion of LNG port facilities. Although technologically and possibly economically feasible, political/social constraints exist. Critics are particularly concerned about the potential for terrorism at LNG port facilities. If targeted by a terrorist group, a large LNG explosion could have a devastating impact on local communities (see Parfomak 2003: CRS-4 –CRS-5).

Uses for natural gas are expanding beyond the traditional thermal energy market. In compressed form, natural gas is used to operate internal combustion engines and produces fewer emissions than gasoline on a per British thermal unit basis. In response to tightened emissions standards, some cities have adopted natural gas as a fuel for mass transportation (Gaul and Young 2003: 12). Natural gas is also used to fire turbines for electricity generation, although its use in the electricity generation market is minimal when compared with coal-, nuclear-, and petroleum-based energy.

Natural gas is a traditional fuel source of growing importance to the U.S. fuel paradigm. Although emissions reductions can be achieved through the use of natural gas, there are two factors that limit long-term status in the fuel paradigm. First, natural gas is not renewable. At some point, availability will fall off and prices will rise significantly. As it is, current demand has driven prices up quite substantially. Second, the exploration of natural gas is technically feasible, but is not always political/socially acceptable. Extraction of coal bed methane involves continued resource extraction in the form of virgin coal, or the reclamation of older mines, which may pose safety risks—again, both plans face forms of political and social opposition. In the long run, other fuel paradigms must be pursued. Currently, there are a number of applications for next-generation fuels.

OTHER ALTERNATIVE FUELS

Hythane™

Hythane™ is a blend of hydrogen and natural gas (typically, methane). Hythane™ is a registered trademark of Brehon Energy PLC. The advantage of Hythane™ is that represents a transitional fuel that can be used in current light-duty vehicles and buses using, for the most part, current generation internal combustion engines (Langreth and Fritz 1994). This means that the use of Hythane™ would not present a significant initial cost for fuel use. Another advantage to Hythane™ is that it significantly reduces emissions in transportation applications (Sierens and Rosseel 2000). According to the manufacturers of Hythane™, nitrous oxide emissions are reduced from 45 to 50 percent as a result of using Hythane™ as opposed to traditional fossil fuels. Finally, Hythane™ requires a small amount of hydrogen in its production, which means that the fuel is less expensive than pure hydrogen. As the hydrogen economy develops, pure hydrogen costs are likely to decline; but, at the moment, pure hydrogen as a fuel source is very expensive (www.hydrogencomponents.com/hythane.html, accessed August 14, 2005).

Clean Diesel

As mentioned earlier, the large-scale use of clean diesel is driven to a large degree by environmental regulation, which requires a significant reduction in sulfur emissions— specifically, a 95 percent reduction in sulfur dioxide emissions from diesel vehicles (see HP Innovation 2005).

While next-generation diesel engine systems are being constructed to burn fuel more efficiently and with reduced emissions (see Ehrenman 2005), existing vehicle fleets will require retrofitting to reduce emissions from existing vehicles. In 2005, the Environmental Protection Agency (EPA) provided $1.6 million in grants for retrofitting projects (see Pollution Engineering 2005). In many states, such as California and Oregon (www.hydrogencomponents .com/hythane.html, accessed August 14, 2005]), clean diesel retrofit programs provide tax credits to owners who retrofit their vehicles. Research on clean diesel has demonstrated that approximately 90 percent of the harmful emissions from traditional diesel engines can be eliminated through clean diesel retrofitting and the sequestration of particulates and the use of fumigation to control NOx emissions (www.energy.ca.gov/afvs/clean_diesel.html, accessed August 14, 2005]).

Biodiesel

Clean diesel research programs have also focused on the development of a product known as biodiesel. The fuel is made from virgin oils from plant seeds such as canola and animal fats. The oil and/or fats are purified and water is removed. The triglycerides present are separated and purified through a process known as transesterification (Meher et al. 2006). It has been shown that blending traditional diesel with biodiesel significantly reduces emissions due to absorption characteristics of compounds found in biodiesel (Duràn et al. 2006). Biodiesel development has been a function of public-private partnerships, involving the DOE, Office of Energy Efficiency and Renewable Energy.

Ethanol

Under EPAct of 1992, a blend of 85 percent ethanol and 15 percent gasoline (E85) is considered an alternative fuel. Ethanol is made from sugar feedstock—often corn, sugar beets, or sugar came. After the feedstock is harvested, it is ground up and the sugar is separated. The sugar is then used as a food supply for microbes. One byproduct produced from microbial digestion is ethanol.

Ethanol raises the level of oxygen in gasoline when used as a blended fuel. Higher oxygen levels contributed to higher performance of the fuel and reduced air emissions, the latter an important requirement contained in the amended Clean Air Act of 1990. While ethanol is one method of oxygenating gasoline, another substance known as methyl tertiary-butyl ether (MTBE) is also used for oxygenation. MTBE can be produced in large quantities at a lower price, but MTBE has been linked to groundwater and soil contamination—an undesirable aspect of the fuel additive. The state of California phased out use of MTBE between 1999 and 2002. A transition to E85 is a long-term goal, but at the moment, the demand for ethanol outstrips supply. Currently, California consumes approximately 900 million gallons of ethanol blended fuel per year in a 5.7 percent blend (Perez 2005; see also Kennedy 2005).[3]

Domestic biomass ethanol production is projected to take several decades to match domestic demand for blended fuel (DiPardo 2002). Federal incentive programs for ethanol production in the Midwest were part of the 2006 U.S. budget to increase domestic supply and reduce demand on foreign energy supplies, primarily the Caribbean and Brazil (Perez 2005). Given high demand, ethanol-blended gasoline has marginally contributed to the rise of gasoline prices.

THE HYDROGEN INITIATIVE

While natural gas remains a centerpiece of the Bush Administration's energy plan, renewable fuels are also being promoted for future use. In his 2003 State of the Union address, President Bush announced a $1.2 billion initiative to develop a hydrogen-based economy. According to the White House, fuel cells are "a proven technology." The initiative contends that technical and economic uncertainties related to the use of hydrogen in fuel cells will be overcome through efforts to lower the cost of hydrogen and fuel cells as well as developing better methods of storage. The initiative intends to link the hydrogen economy to President Bush's "Freedom Car" initiative, which establishes the goal of energy independence by 2020. The initiative concludes that the development of the hydrogen economy will reduce air pollution and greenhouse gas emissions by "more than 500 million metric tons of carbon equivalent each year by 2040" (White House 2003).

The plan was criticized as being built more on false promises and without full recognition of the difficulties facing the development of a hydrogen economy (Hoffman and Harkin 2003). Many difficulties facing alternative energy policy may have been made worse because of a lack of commitment to previous policy goals and overreliance on fossil fuel and existing energy infrastructure (see Jaskow 2003; Jorgenson 1981; Mattoon 2002). By contrast, a more steady European commitment to green energy and environmental standards (see Dunn and Flavin 2002) has produced more coherent energy policy results in the European Union (see Miller et al. 2002). The initiative may be overly optimistic in its timeline to develop "sound science" (see Moritsugu 2002) and the promise of independence of foreign oil supplies (see Schroeder 2002; Wald 2004). Nevertheless, proponents such as Jeremy Rifkin (2002, 2003) conclude that a hydrogen-based economy will come to fruition (see also Lakely [2003] regarding Secretary of Energy Spencer Abraham).

Some concerns exist about the plan's top-down policy approach and its potential conflict with President Clinton's 1990s initiatives to create sustainable cities. As has been pointed out in the extant literature, sustainable cities/ communities require a local-level, bottom-up approach to policy. Communities must develop a sense of realistic growth balanced with resource needs (see Nijkamp and Pepping 1998; Portney 2002). By focusing on the local level, communities will make wiser choices regarding growth feasibility, standards of living, and energy prospects (Staka 2002). But, as Vignola et al. (2000) illustrate in their analysis of photovoltaic (PV) systems in public schools, a local sense of energy needs must be balanced with sound science that is established at appropriately funded levels of government. In order for local needs to be balanced with sound science, local stakeholders must develop an understanding of energy capacity, which requires that citizen decision makers understand technical information in relation to community needs. With these considerations in mind, the Hydrogen Initiative compliments the sustainable communities model, promoting understanding and acceptance of a new energy paradigm.

HYDROGEN AND FUEL CELLS

Technical Feasibility Issues

The major advantage to the fuel cell is its efficiency in producing electricity. Hydrogen fuel cell development can be traced to studies conducted by Sir William Grove in 1839, whose early work was refined by Francis Bacon in 1932 (Francis 2002: 34; Ketelaar 1993). The basic idea behind the fuel cell is that pure hydrogen is isolated from oxygen molecules within a fuel cell system. In a fuel cell, an anode attracts the electrons—thus forming H^+ ions— which are then free to travel though electric circuits. The cathode or positive end of the fuel cell distributes oxygen molecules along its surface. A proton exchange membrane (PEM) dividing the fuel cell anode and cathode allows

H+ ions to travel freely to the cathode, but restricts the passage of electrons, which are then forced through electric circuits. A catalyst composed of platinum powder facilitates the combination of the H+ ions, O_2 molecules, and the electrons traveling to the cathode; the resulting combination is H_2O or pure water.

Pure hydrogen is a highly explosive gas commonly thought to be dangerous. The evidence of the gas's explosive qualities is mixed (Vaitheeswaran 2004). In its favor, hydrogen is an odorless gas and not harmful to humans except in very high concentrations. The gas also rises and disperses into the atmosphere very quickly, which reduces its threat in outdoor applications. Hydrogen gas also has a high burn/explosive velocity, which means that it burns very quickly and is less explosive as a gas (see Rigas and Sklavounos 2005). Based on these positive qualities of hydrogen, many technical proponents of fuel cell vehicles argue that it is a safe energy source for vehicle use and for energy storage.

However, technically-minded critics point to several areas of concern with regard to hydrogen fuel cell vehicles. First, hydrogen fuel cells have the potential to run on very high voltages or rates at which electrons pass through circuits. Exposure to voltages of greater than 50V can cause heart failure—some fuel cell vehicles run on 350+V, which increases the risk of electrocution (College of the Desert 2001). A second area of concern is the inability to rapidly detect failures in the fuel cell. Tanrioven and Alam (2005) assessed the reliability of grid-connected PEM fuel cell (PEMFC) power plants. Although not a transportation system, the PEMFC system requires auxiliary power generation systems to maintain generation capability due to derated state or system failures. PEMFC system operations are impacted by hours of operation and age of system. Failure rates become much higher after eight years of operation. Temperature variation impacts in winter months impact PEMFC to small degree (Tanrioven and Alam 2005: 277).

Production and storage of hydrogen are additional areas of technical problems and conflicting conclusions. Hydrogen is the most abundant element in the universe, but is not easy to produce and store for use in energy systems. On Earth, hydrogen is not found in its elemental state, but rather forms chemical bonds with other elements. All petroleum-based products contain hydrogen, but the energy inputs required to remove hydrogen is extremely high and result in carbon emissions. The hydrogen for fuel cell use will be "clean" energy, but the organic source material must be considered, too.

Green power approaches are second methods of producing hydrogen. In essence, hydrogen would be produced from water through a process known as electrolysis. The use of fossil fuels to run electrolysis plants would make little technical sense because, in essence, there would be a huge loss of energy in the hydrogen production process (i.e., more energy would be lost to create hydrogen gas than would be produced from the resulting hydrogen gas used in fuel cell vehicles). Therefore, the use of wind, solar, or geothermal energy to produce electricity to be used to operate electrolyzers for the production of hydrogen is the basis of the green power approach to producing a net gain in energy production.

Several issues must be addressed in the green power model. First, the model requires a tremendous capital investment in alternative energy production systems. A Vestas V-80 wind turbine—a well-known Danish manufacturer—has a listed cost of $1.35 million (Devine et al. 2003; Herzik et al. 2004). Wind turbine prices are highly relative to demand, consumer credit ratings, and government price credits (Wustenhagen 2003). Share prices for producers have experienced some turmoil, which impacts the capacity of corporations to raise capital through stock issuances. PV technology is in a period of transition; systems are becoming cheaper but still remain very expensive. The operations and maintenance costs associated with wind and solar systems varies on location, climatic conditions, and use. Additionally, tremendous investment must be made in battery storage units to store power for use during times when there is no wind or the weather is suboptimal. Batteries contain acid and lead, which are known to be harmful to the environment; also, batteries must be replaced and dissembled for rebuilding, which is an additional environmental hazard. The system will also require substantial land for system installation. The NIMBY ("not-in-my-back-yard") concerns with wind turbines and solar panels are fairly substantial.

Storage of hydrogen gas is also a source of conflict on the technical feasibility side of hydrogen fuel cell use. The common method of storing hydrogen is in a gaseous, liquid form. Gaseous hydrogen is stored in tanks under pressure. Because hydrogen is the smallest element, it tends to be an "escape artist," and there is a potential for fuel losses in storage and transportation. Storage tanks must be heavily lined with lead sheathing to reduce gas losses. Another method of storing hydrogen is in a super-cooled or liquid form. The process of liquefying hydrogen is very expensive because of the tremendous pressure and the need for maintaining a super-cooled state to avoid tank rupture. Solid hydrogen is actually a gaseous state in which the hydrogen molecules bond with metal hydride filigrees constructed within tanks (Zuttel 2003). In its bonded state, hydrogen must be returned to a free gaseous state for use in a fuel cell. On a final note, the storage of hydrogen is further hampered by the gas's ability to weaken metal structures, which may cause gas losses as well as safety issues.

Geothermal power can be used to produce electricity for the production of hydrogen. There are a limited number of locations in the United States and elsewhere that could be developed for geothermal power—in the United States, the southwestern states hold the most promise for geothermal development. There are no emissions from geothermal power generation, which would make the hydrogen production process very "green" (Penner 2006).

Economic Feasibility Issues

Economic feasibility of large-scale hydrogen production is not clear at this time. Economic feasibility quandaries can be categorized into at least three different dimensions. First, cost and benefit considerations are vague. Costs are often unclear due to continuing cost declines in the area of renewable energy (in terms of PV market predictions, see Garvison 2003; Pryor and Wilmot 2001; UNEP 2003). Wind energy capital costs have come down from $2,500/kWh in the early 1980s to less than $1,000kWh in 2003 (Reeves 2003: 10). Capital costs are coming down, but the recurring costs associated with operations and maintenance (O & M) of renewable energy systems remain unclear (see Moore et al. 2003: 416), particularly for large-scale renewable energy operations. Existing large-scale systems remain hesitant to share their O & M costs either because the high costs of O & M would be politically damaging or because the O & M costs are proprietary information that could be used by competitor industries. Using small-scale cost figures (e.g., Vignola et al. 2000) is not reliable because the figures may not be linear in terms of cost scalability. Finally, O & M costs are largely a function of wear/use factors and, in many cases, are unscheduled maintenance events (see Reeves 2003: 10).

In other ways, costs are difficult to place solely in monetary terms. For example, building a large-scale wind farm on a pristine hillside has some effect on the physical beauty of the landscape; is it possible to measure that cost in a reliable, valid, and political and socially acceptable way? The development of alternative energy resources, therefore, impose many variable costs that remain "hidden" until that time when pressure group information campaigns or NIMBY activists reveal these costs, in many cases seeking to identify a cost figure. Economic feasibility is driven to a significant degree by political feasibility demonstrated through policy initiative and public acceptance (Maack and Skulason 2006).

Operations and maintenance costs are often simultaneously used as measures of cost and benefits. Alternative energy publications will often include operations and maintenance in estimates of local economic development resulting from the use of alternative energy sources such as wind and photovoltaic systems. As wind and PV systems reports indicate, many costs associated with O & M are fixed costs, which could be viewed as the basis of local economic development, employing a steady workforce to maintain alternative energy systems. Conversely, other reports indicate that a sizable portion of the O & M costs for alternative energy systems are variable O & M costs; unscheduled maintenance of energy production systems, which would not be a steady source of economic development. In the case of variable O & M costs, specialized repairs to wind or PV systems needed in large-scale green hydrogen production, for instance, would likely involve that manufacturer-provided technologists be brought to the site.

Benefits are a function of how alternative energy sources are developed. For instance, if hydrogen is produced from natural gas, then the resulting carbon must be managed in a manner consistent with existing national environmental policies. Emitting carbon into the air would be a cost, although proponents argue that the use of the hydrogen in fuel cell vehicles is a benefit.

In order to avoid the cost debate, many scholars advocate carbon sequestration. One method of storing sequestering carbon calls for pumping carbon into ocean beds or into deep Earth mine shaft or depleted oil wells. The simple laws of thermodynamics, however, would indicate that matter, sequestered or not, cannot be destroyed. Therefore, the cost debate is really a matter of "where" the carbon will be after hydrogen is stripped from natural gas. Proponents of green energy argue that this debate is moot if one produces hydrogen using electrolysis. Skeptics point out that there is still much work to be done before green power can fully supplant the use of fossil energy for electricity generation—an absolute requirement for the mass production of green hydrogen.

Policy analysts often use "multipliers" to determine the economic development benefits that will emerge from the development of new industries in a local community. Alternative energy economics literature is slowly coming to terms with appropriate multipliers for green power industries. A recent report produced by scholars at the University of Nevada, Las Vegas, has identified "high use" and "low use" multiplier estimate for renewable energy production of 1.72 and 1.67 respectively (Schwer and Riddel 2004); notably, the authors do not factor in the benefits to air quality.

The economic information quandary facing alternative energy systems is a lack of systematic analyses of costs and benefits. A clear determination of what costs and benefits entail has yet to be uniformly measured.[4] The problem with this, of course, is that absent uniform measure of costs and benefits, it is very difficult to make economically wise decisions as a producer and as a consumer.

OTHER TYPES OF FUEL CELLS CURRENTLY IN USE AND/OR DEVELOPMENT

Alkaline Fuel Cells

Alkaline fuel cells (AFCs) were developed in the 1960s and used aboard spacecraft in the Apollo missions as a source of power. AFCs are also used aboard space shuttle vehicles. Given the energy creation process, an AFC can also produce potable water, which is a valuable asset in space flight. The inputs into an AFC fuel cell are H_2 and O_2. The former input occurs on the anode end of the fuel cell, while the latter is on the cathode end of the fuel cell. It should be noted that both of these materials are potential fire hazards, therefore the AFC must be carefully managed to protect against possible accidents.

The electrolytic material between the anode and cathode portion of the fuel cell is composed of potassium hydroxide. Potassium is part of the alkalai earth metals family of elements and has a tendency to give up an electron needed by an OH^- molecule, thus forming an ionic chemical bond. Visualizing the electrolytic material, one would likely see it in the form of a white flaky material called caustic potash or lye. The material is easily dissolved in water and gives off tremendous heat when that occurs—this means that it has exothermic qualities. Potassium hydroxide also "likes" to absorb CO_2, which would effectively destroy its value as an electrolyte in the AFC fuel cell. The system must be closed so to prevent CO_2 from corrupting the fuel cell (note: this is much easier to do in outer space than in terrestrial applications).

The fuel cell operation is not dissimilar from the hydrogen fuel cell system discussed later in this chapter. In essence, at the anode end, covalently bonded hydrogen molecules are attracted to the hydroxyl ions in the electrolyte, producing water and four free electrons. The electrons travel to the cathode side of the fuel cell via the electrical circuits (where they can be used as an energy source). At the cathode end of the fuel cell, the electrons chemically interact with oxygen molecules and water molecules, producing four hydroxyl ions (OH-) that are then draw into the electrolytic portion of the fuel cell: the cycle begins anew (see Mugerwa and Blomen 1993 [www.fctec.com, accessed August 13, 2005]).

Direct Methanol Fuel Cells

Direct methanol fuel cells (DMFCs) were first developed in the 1990s and operate similarly to standard hydrogen fuel cells. At the anode end of the cell, liquid methanol is introduced to pure water, producing CO_2 gas, hydrogen ions, and electrons. The hydrogen ions travel through the electrolyte but the electrons are forced to travel through an electrical circuit to react, on the cathode end of the fuel cell, with oxygen and the hydrogen ions to form water. DMFCs remain in development stages with low efficiency rates at the present time (<10 percent). The polymer materials required for the process are a bit more expensive because they must be sturdier in the chemical reaction process than the thinner platinum used in PEM hydrogen fuel cells; this makes the DMFC option more expensive currently. It is hoped that with continued research and development, DMFCs could eventually be used to power automobiles and other consumer products.

Molten Carbonate Fuel Cells

Molten carbonate fuel cells (MCFCs) were first developed in the 1950s by two Dutch scientists named G. H. J. Broers and J. A. A. Ketelaar (www.fossil .energy.gov/programs/powersystems/fuelcells/fuelcells_moltencarb.html, accessed August 14, 2005). In the 1960s, Texas Instruments developed small-scale (i.e., 100W–1kW) MCFCs for the U.S. Army. Following the oil shocks of the late 1960s and early 1970s, the DOE formed cooperative relationships with private sector enterprises (e.g., FuelCell Energy) to promote research, development, and production of MCFCs. According to the Office of Fuel Energy, DOE, FuelCell Energy MCFC units "are operating at 42 installations nationwide" (www.fossil.energy.gov/programs/powersystems/fuelcells/fuelcells_ moltencarb.html, accessed August 14, 2005]).

MCFCs require tremendous energy inputs to operate. As the name of the fuel cell would imply, the process requires melting an electrolytic material known as lithium potassium carbonate salts, requiring temperatures at or exceeding 1,200° F. In a molten state, the electrolyte is able to attract ions that are passed between the anode and cathode. The MCFC fuel cell anode input is hydrogen (H_2), which interacts with the molten electrolyte to produce water, carbon dioxide, and two free electrons that travel through electrical circuitry (available for use as a source of power) to the cathode. The inputs at the cathode are oxygen (O_2) and carbon dioxide (CO_2). The carbon dioxide produced by the anode reaction can be recaptured and become an input for the cathode, as CO_2 levels need to be replenished in the electrolyte due to the loss of carbon dioxide at the anode end chemical reaction (Anahara 1993).

The MCFC requires and produces tremendous thermal energy. The high temperature operation means that chemical compounds of which the fuel is comprised move more freely and therefore the efficiency of electricity production is much higher. If the thermal energy used and produced in the MCFC fuel cell is recaptured and used to keep the fuel cell process operating, the production efficiency of the fuel cell "can be as high as 85 percent" (www

.fossil.energy.gov/programs/powersystems/fuelcells/fuelcells_moltencarb .html, accessed August 14, 2005).

A major disadvantage of MCFC cells is that the very high temperature at which they operate restricts the variety of applications—for instance, it would not be safe in residential applications (www.visionengineer.com/env/mc .shtml, accessed August 14, 2005). Another disadvantage to MCFCs is that high temperature production units have a greater tendency toward corrosion and system failure; thus, the operation and maintenance of MCFC units is likely to be relatively expensive (Roberge 2005 [www.corrosion-doctors.org/ FuelCell/mcfc.htm, accessed August 14, 2005]).

Currently, FuelCell Energy is working on a 3MW plant design. Another advantage to MCFCs is that the system does not require the expensive platinum catalysts required in other fuel cell systems; the Office of Fossil Energy reports that nickel catalysts are sufficient for MCFC units (www.fossil.energy .gov/programs/powersystems/fuelcells/fuelcells_moltenc arb.html, accessed August 14, 2005) According to Roberge (2005): "If built in low numbers, MCFCs are likely to cost around $3,000/kW. If costs can be reduced to $1,500/kW, which would require order commitments to support high-volume manufacturing, these systems could find significant utility markets for distributed generation in grid-support applications."

Phosphoric Acid Fuel Cells

Phosphoric acid fuel cells (PAFCs) were first developed in the 1960s and 1970s. The basic design of a PAFC is not dissimilar from other fuel cells discussed. The fuel cell operates at optimum levels at temperatures ranging between 300° and 400° F. As with other fuel cell systems, PACFs usually use platinum catalysts at both the anode and cathode. Hydrogen is the input at the anode. Hydrogen ions migrate through the phosphoric acid electrolyte, while the freed hydrogen electrons travel through electrical circuits to perform useful work. The input at the cathode is oxygen, which can be ambient air from the atmosphere. Experiments have shown that the carbon dioxide in ambient air will be tolerated by a solid oxide fuel cell (SOFC) up to a certain concentration level. The output at the cathode is pure water (Van den Broek 1993).

The operating efficiency of PAFCs is approximately 40 percent, but if used in a "cogeneration" system—heat input sources coming from excess heat generation at fossil fuel fired energy plants—the efficiency rate can be as high as 85 percent (FCTec 2005 [www.fctec.com/fctec_types_pafc.asp, accessed August 14, 2005]). One of the advantages to PAFCs is that they are very useful in small applications, such as homes and commercial enterprises. Running at cooler temperatures, PAFCs are safer systems to operate. Nevertheless, the price of a PAFC is still quite substantial and would likely require incentive programs to make the systems affordable to the average homeowner or business. According to the Office of Fossil Energy, DOE, the typical PAFC costs between $4,000 and $4,500 per kW plant size.

Solid Oxide Fuel Cells

SOFCs were first developed in the 1930s by Swiss scientists. Despite many developmental problems discovered by European and Russian scientists in the 1940s and 1950s, SOFC research was not discontinued. In the early 1960s, the Westinghouse Corporation conducted research on the development of a viable SOFC system.

SOFC systems run at very high temperatures (approximately 1,800° F). The electrolytic portion of the SOFC is composed of a series of thin tubes. The tubes often resemble a compact diskette in terms of their thickness (chemelab.ucsd.edu/ fuelcell/soxide.htm, accessed August 14, 2005), and are often hollow, allowing air to flow through the tube. Along the outer portion of the inside of the tube, a solid oxide compound is applied. Studies conducted by Westinghouse used zirconium oxide and calcium oxide. The chemical reactions that occur may require methane as an input on the anode portion and ambient air as an input at the cathode, although oxygen molecules are the primary reactive elements needed. The process produces carbon dioxide and water as outputs, as well as substantial electrical power (chemelab .ucsd.edu/fuelcell/soxide.htm, accessed August 14, 2005). Other systems may use pure hydrogen gas an anode input, producing only pure water and electricity as an output. One significant advantage to the SOFC is that it is often a ceramic-type tubing, which is more resistant to corrosion problems brought on by the extreme operational temperatures (see also Murugesamoorthi et al. 1993).

APPLICATIONS FOR NEXT GENERATION
ALTERNATIVE FUELS, FUEL CELLS

Applications for alternative fuels—beyond the natural gas paradigm—has been primarily initiated by government regulations and clean air benchmarks in domestic regulations and in international agreements such as the Kyoto Treaty. The United States is not a signatory to the treaty, but the goals on which it is grounded have been offered tacit, albeit qualified, support by the George W. Bush administration. Briefly discussed earlier, ethanol (manufactured from corn) is a form of alternative fuel that can be used in gasoline-powered engines. Ethanol has been a subsidized energy product for several decades and has been touted as a cleaner renewable energy source of the future. Clean diesel and biodiesel applications are of immediate importance in alternative fuel development. The EPA has developed a reduced sulfur rule for diesel that will require a 95 percent reduction in sulfur emissions for trucks produced in the 2007 model year and beyond (Milbourn 2001). Hythane™ is also a product that has been developed that blends natural gas with hydrogen to produce a cleaner burning fuel that requires a smaller hydrogen fuel production infrastructure because it requires less hydrogen.

In terms of futuristic fuel sources, fuel cell vehicles, primarily hydrogen fuel cells, are being promoted through state legislation, such as California's Hydrogen Highway initiative (Baker and Jollie 2005: 11) as well as cooperative initiatives among many upper midwestern states in the United States. Demand via government regulation is also driving the development of alternative fuels and fuel cell use in Europe (via the Clean Urban Transport for Europe—CUTE—Program).

Transportation

Transportation is one of the largest consumption sectors in the fuel market. According to the Energy Information Agency, DOE, 40 percent of petroleum demanded in the U.S. market is sold and used in the form of gasoline for transportation, and 90 percent of all petroleum used in the United States is in the form of motor fuels (see Energy Information Agency 2005 [http://www.eia.doe.gov/neic/infosheets/petroleumproducts.htm, accessed August 8, 2005]). Of some note, given the exceptionally high petroleum prices in nominal terms in the current market, petroleum demand has not substantially declined. The mobility and sustainability of the modern society or community is imperative; therefore, safe, reliable, and effective fuel alternatives must be researched and developed.

While the natural gas model has become more prevalent, another simple transition involving next generation fuel would involve a clean diesel vehicle. In other words, clean diesel engines would replace the current generation of internal combustion gasoline engines, in part driven by aforementioned EPA standards. Of the two requisites outlined by many proponents of alternative fuel, clean diesel addressed only one issue: namely, the issue of air-, water-, and soil-borne fuel emissions. It would not, however, move the fuel paradigm beyond reliance on fossil fuels.

Alternative fuels such as hydrogen for fuel cells is also being developed for large-scale use. Hydrogen fuel cells are being developed in mass transit bus systems and are currently either used or tested in many metropolitan areas in

Europe, Canada, the United States, and elsewhere. Hydrogen fuel cells have demonstrated their capacity to produce the large quantities of power needed to move a bus at an efficient speed. Approximately sixty-five fuel cell buses are in use worldwide, with projections showing that this number will dramatically increase over the remainder of the decade and beyond (see Adamson 2005: 1).

> Bus fleets are being seen as a good early market for fuel cell and hydrogen technology for reasons such as central refueling, predefined routes (so operators can calculate time between refueling and distances), high public visibility, size, weight of the stack not being as critical as in the light duty vehicle market, and more design space for hydrogen storage tanks. [Adamson 2005: 4]

Light-duty utility vehicles face the primary problem of trying to gain access to a refueling site. The infrastructure for hydrogen refueling is being developed, but projections for a refueling infrastructure, as with vehicle development itself, are being pushed back by manufacturers to perhaps 2020.

In a 2005 study of transportation and fuel cell technology, niche transportation units are growing in popularity. Niche transportation units are not primary vehicle systems, but are an important first step in the development of affordable, reliable, easily refueled personal and mass transportation.

In the 1960s, the market demand tended toward the development of scooter transportation, forklifts, and various forms of marine/submarine applications (Adamson 2005: 2). What is particularly interesting is that while the National Aeronautics and Space Administration has used fuel cells in spacecraft, the survey of niche market demand presented here demonstrates that aerospace applications have never been a significantly large portion of the niche fuel cell transportation market. By 2005, there was a dramatic shift in market demand (see Adamson 2005: 2). Scooter and forklift transportation demand have declined significantly, while alternative power unit and marine-related uses have increased quite substantially.

Alternative power units (APUs) are often used in the trucking industry for basic operations of a truck and its cargo maintenance while the truck is stopped for refueling or for operator rest periods. Given the price of diesel in the current market, it is not efficient to operate the main diesel engine during these periods when the truck is not moving; APUs substantially reduce variable costs of operation and maintenance. Maritime uses tend to be focused on military applications and are not widely known, but operating system power for basic energy requirements (e.g., water treatment plant, lighting, etc.) rather than main propulsion are my guess of the likely uses for submarine niche transportation fuel cell systems (see also Baker and Jollie 2005). Miscellaneous applications for wheelchairs and other disability-associated transportation niches is also, at the very least, a stable area of demand for fuel cell systems. Finally, some progress is being made in domestic aerospace companies such as Boeing, as well as internationally recognized aerospace firms, to develop fuel cell systems to operate aboard aircraft, likely serving as APUs systems at first, but possibly used for primary or secondary sources of energy for thrust-related systems.

The development of fuel cell systems for niche market demand is growing so rapidly that it is very difficult to remain current on market changes. The cumulative number of niche transport systems more than doubled between 2004 and 2005. Market distribution has changed substantially in only one year, too. In the period 1963–2004, the majority of niche transportation fuel cell systems were distributed in Europe. By 2005, North America had easily replaced Europe in terms of units distributed. At the moment, market demand is still developing, likely due, in part, to changing governmental regulations and incentives.

Military Applications

Military applications go beyond submarine-related systems and may not be well reflected in the Adamson's (2005) study, largely because in terms of systems development, many of the military applications remain in research and development stages. There are several examples of military applications that illustrate the flexible use of alternative fuel cell systems. One system developed by Teledyne provides energy as well as fresh drinking water to flight crews. The obvious value in this system would be that military aircraft could sustain longer missions (Baker and Jollie 2005: 2). The U.S. Air Force is developing a very high altitude helium dirigible system with fuel cell energy technology providing up to 500 kW_e to operate in-flight surveillance equipment. AeroVironment has developed a High Altitude Long Endurance pilotless aircraft that operates on fuel cell technology and that is capable of remaining airborne for a week (Adamson 2005: 6), which has obvious military applications for surveillance operations; however, the system remains in development stages. Batelle is developing an auxiliary power unit for Bradley fighting vehicles. The Quantum Corporation is developing the "Aggressor," a battlefield transportation system fueled by hydrogen fuel cell technology that is capable of acceleration to forty miles per hour. Many portable battlefield devices are under development to power fighting equipment, as well as the computer equipment, frequently carried by soldiers on the modern battlefield. Soldiers use computer equipment for navigation, air traffic management and air strike requests, and general communication. The plurality of systems developed (36 percent) are related to portable device fuel cell energy systems. Weapons systems account for less than 10 percent of resources directed toward application of fuel cell technology for military purposes.

The United States and Canada are taking the lead in developing uses for fuel cells in military operations. According to Baker and Jollie (2005: 5), over 70 percent of the organizations involved in military fuel cell development are based in North America. European allies account for less than a quarter of all organizations involved in military fuel cell application development. Currently, Middle Eastern and Asian organizations account for less than 10 percent of the military applications of fuel cells.

Residential

Residential uses for alternative fuels often focus on residents seeking methods of storing green energy that has been produced by the homeowner. Green energy can be net metered or sold back to the power grid.[5] There are instances when a net-metering option is neither available nor desirable. For instance, an individual homeowner may live some distance from power lines and may wish to store energy for his or her future use. A traditional method of storing energy in these circumstances is using lead-acid batteries, but batteries of this type are not environmentally friendly and require special venting, regular maintenance, and replacement. Energy could also be stored in the form of hydrogen, produced in the same method described earlier but on a much smaller scale.

Commercial

Businesses that manufacture and sell products and services often own factories or retail spaces. As noted earlier, auxiliary power units are one of the fastest-growing areas for fuel cell technology demand. Commercial enterprises, particularly production units, require easily transportable and reliable energy sources. Fuel cell technology can easily provide such benefits.

CHAPTER SUMMARY

This chapter has focused on a handful of fuels and fuel systems that are either currently being employed or in developmental stages. Obviously, there are several other types of fuel that could be discussed and greater attention to scientific details could be presented. But the purpose of the chapter was to provide a general overview to the fuels that are available as alternatives to fossil energy and that will likely be used in the twenty-first century as basic components in sustainable communities.

NOTES

1. In proportion to other fuel sources, wood has declined in use. Proportionally, coal use peaked in the early twentieth century (Podobnik 2006: 5).
2. Parfomak (2003) indicates that seven additional LNG off-loading sites are pending approval by the U.S. government.
3. Pat Perez is an analyst for the California Energy Commission.
4. The economic information quandary is a good example of Deborah Stone's (1988) analysis of the meaning of statistics in public policy analysis—numbers have meanings assigned both by analysts and other critical consumers of information.
5. This form of energy production and net metering of surplus energy is part of the distributed energy production paradigm.

WORKS CITED

Adamson, K. 2005. Fuel Cell Today Market Survey: Niche Transport (Part I). *Fuel Cell Today*. www.fuelcelltoday.com, accessed August 11, 2005.

Anahara, R. 1993. Research, Development, and Demonstration of Molten Carbonate Fuel Cell Systems," in Blomen, L. and Mugerwa, M., eds. *Fuel Cell Systems*. New York: Plenum Press, pp. 271–343.

Ashtiani, C., Wright, R., and Hunt, G. 2006. Ultracapacitors for Automobile Applications. *Journal of Power Sources* 154(2): 561–66.

Baker, A. and Jollie, D. 2005. Fuel Cell Market Survey: Military Applications. *Fuel Cell Today*, www.fuelcelltoday.com, accessed August 11, 2005.

College of the Desert. 2001. *Module 6: Fuel Cell Engine Safety*. www.eere.energy.gov/hydrogenandfuelcells/tech_validation/pdfs/fcm06r0.pdf, accessed November 27, 2005.

Claycomb, J., Brazdeikis, A., Le, M. Yarbrough, R., Gogoshin, G., and Miller, J. 2003. Nondestructive testing of PEM fuel cells, *IEEE Transactions on Applied Superconductivity* 13(2): 211–14.

Devine, M., O'Connor, B., Ellis, T., Rogers, T., Wright, S., and Manwell, J. 2003. *Massachusetts Wind Energy Predevelopment Support Program and Feasibility Study for Marblehead, Massachusetts*. Amherst, MA: Renewable Energy Research Lab, University of Massachusetts. www.ceere.org/rerl/publications/reports/WEPS_and_Marblehead_Wind_Feasibility_AWEA03.pdf, accessed November 27, 2005.

DiPardo, J. 2002. Outlook for Biomass Ethanol Production and Demand [working paper last modified July 30, 2002]. www.eia.doe.gov/oiaf/analysispaper/biomass.html, accessed March 29, 2006.

Dunn, S. and Flavin, C. 2002. The Climate Change Agenda: From Rio to Jo'burg and Beyond. *International Journal of Technology Management & Sustainable Development* 1(2): 87–111.

Duràn, A., Monteagudo, J., Armas, O., and Hernàndez, J. 2006. Scrubbing Effect on Diesel Particulate Matter from Transesterified Waste Oils Blends. *Fuel* 85(7/8): 923–28.

Ehrenman, G. 2005. Cleaner Fuel Economy. *Mechanical Engineering* 127(5): 12–14.

Energy Information Agency. 2005. Petroleum Products. www.eia.doe.gov/neic /infosheets/petroleumproducts.htm, accessed August 8, 2005.

FCTec. 2005. Fuel Cell Basics. *Fuel Cell Technologies.* www.fctec.com/fctec_about .asp, accessed August 14, 2005.

Francis, M. 2002. Modeling: Driving Fuel Cells. *Materials Today* 5(5): 34–39.

Garvison, P. 2003. Solar Markets and Storage. *Systems-Driven Approach for Solar Applications of Energy Storage.* Washington, D.C.: U.S. Department of Energy.

Gaul, D. and Young, L. 2003. U.S. LNG Markets and Uses. *Energy Information Administration, Office of Oil and Gas,* January. www.eia.doe.gov/pub/oil_gas/natural _gas/feature_articles/2003/lng/lng2003.pdf, accessed March 20, 2006.

HP Innovation. 2005. Process Converts Oil Sands Bitumen into Low-Sulfur Distillates. *Hydrocarbon Processing* 84(7): 30.

Herzik, E., Simon, C., and Marks, S. 2004. Economic Feasibility—Life Cycle Cost Study: Regional Transportation Commission, Hydrogen Fuel Project. Reno: T3/University of Nevada.

Hoffmann, P. and Harkin, T. 2003. *Tomorrow's Energy: Hydrogen, Fuel Cells, and the Prospects of a Cleaner Planet.* Boston: MIT Press.

Jacoby, M. 2006. Boost for Battery Performance. *Chemical & Engineering News* 84(8): 10.

Jaskow, P. 2003. Energy Policies and Their Consequences after 25 Years. *The Energy Journal* 24(4): 17–49.

Jorgensen, J. 1981. Social Impact Assessments and Energy Developments. *Policy Studies Review* 1(1): 66–86.

Kennedy, R. 2005. *Ethanol Market Outlook for California,* CEC-600-2005-037. Sacramento: California Energy Commission.

Ketelaar, J. 1993. History, in Blomen, L. and Mugerwa, M., eds. *Fuel Cell Systems.* New York: Plenum Press, pp. 19–36.

Lakely, J. 2003. Abraham Outlines Plans for Hydrogen Fuel, Canadian Oil. *The Washington Times,* November 17, A11.

Langreth, R. and Fritz, S. 1994. Hydrogen + Natural Gas=Hythane. *Popular Science* 244(3): 34–35.

Maack, M. and Skulason, J. 2006. Implementing the Hydrogen Economy. *Journal of Cleaner Production* 14(1): 52–64.

Mattoon, R. 2002. The Electricity System at the Crossroads. *Society*, November/December: 64–79.

Meher, L.,Vidya-Sagar, D., and Naik, S. 2006. Technical Aspects of Biodiesel Production by Transesterification—A Review. *Renewable and Sustainable Energy Reviews* 10(3): 248–68.

Mibourn, C. 2001. EPA Gives the Green Light on Diesel-Sulfur Rule. February 28, *United States Environmental Protection Agency.* yosemite.epa.gov/opa/admpress .nsf/b1ab9f485b098972852562e7004dc686/0237f756e256922c85256a010072e 4f6?OpenDocument, accessed August 12, 2005.

Miller, S., Brushan, B., and J. Ball 2002. "A Global Report: Europe Launches Hydrogen Initiative," *Wall Street Journal*, October 16.

Moore, L., Malcynski, L., Strachan, J., and Post, H. 2003. Lifecycle Cost Assessment of Fielded Photovoltaic Systems. *NCPV and Solar Program Review Meeting—National Renewable Energy Laboratories.* NREL/CD-520-33586, 416–18.

Moritsugu, K. 2002. Hydrogen Fuel Cell Technology Remains Many Years in Future. *Pittsburgh Post-Gazette Journal*, January 13, A10.

Mugerwa, M. and Blomen, L. 1993. Research, Development, and Demonstration of Alkaline Fuel Cell Systems, in Blomen, L. and Mugerwa, M., eds. *Fuel Cell Systems.* New York: Plenum Press, pp. 531–64.

Murugesamoorthi, K., Srinivasan, S., and Appleby, A. 1993. Research, Development, and Demonstration of Solid Polymer Fuel Cell Systems, in Blomen, L. and Mugerwa, M., eds. *Fuel Cell Systems.* New York: Plenum Press, pp. 465–92.

Nijkamp, P. and Pepping, G. 1998. A Meta-Analytical Evaluation of Sustainable City Initiatives. *Urban Studies* 35(9): 1481–1500.

Parfomak, P. 2003. *Liquefied Natural Gas (LNG) Infrastructure Security: Background and Issues for Congress*, CRS Order Code RL32073. Washington, D.C.: Congressional Research Service.

Penner, S. 2006. Steps toward the Hydrogen Economy. *Energy* 31: 33–43.

Perez, P. 2005. Ethanol in California. *Platts Ethanol Finance & Investment Conference*, May 25–26, 2005.

Podobnik, B. 2006. *Global Energy Shifts: Fostering Sustainability in a Turbulent Age.* Philadelphia: Temple University Press.

Pollution Engineering. 2005. EPA Retrofits Diesel Engines. *Pollution Engineering*, April, pp. 10–11.

Portney, K. 2002. Taking Sustainable Cities Seriously: A Comparative Analysis of Twenty-Four U.S. Cities. *Local Environment* 7(4): 363–80.

Pryor, T. and Wilmot, N. 2001. *The Effect of PV Array Size and Battery Size on the Economics of PV/Diesel/Battery Hybrid RAPS Systems.* Murdoch, WA: Murdoch University Energy Research Institute.

Reeves, A. 2003. *Wind Energy for Electric Power: A REPP Issue Brief.* Washington, D.C.: Renewable Energy Policy Project.

Rifkin, J. 2002. *The Hydrogen Economy: The Creation of the Worldwide Energy Web and the Redistribution of Power on Earth.* New York: Putnam Press.

Rifkin, J. 2003. Thinking Big: The Forever Fuel the New Hydrogen Economy Will Not Only Eliminate Our Dependence on Foreign Oil, It Will Turn Our Automobiles into Power Plants. *The Boston Globe*, February 23, D12.

Rigas, F. and Sklavounos, S. 2005. Evaluation of Hazards Associated with Hydrogen Storage Facilities. *International Journal of Hydrogen Energy* 30(13/14): 1501–10.

Roberge, P. R. 2005. Molten Carbonate Fuel Cells. *CorrosionDoctors.Org,* www.corrosion-doctors.org/FuelCell/mcfc.htm, accessed August 14, 2005.

Schroeder, W. 2002. Clear Thinking about the Hydrogen Economy. *Connecticut Law Tribune*, December 20, p. 5.

Schwer, R. and Riddel, M. 2004. *The Potential Economic Impact of Constructing and Operating Solar Power Generation Facilities in Nevada*, NREL/SR-550-35037. Golden, CO: National Renewable Energy Laboratory.

Sierens, R. and Rosseel, E. 2000. Variable Composition Hydrogen/Natural Gas Mixtures for Increased Energy Efficiency and Decreased Emissions. *Journal of Engineering for Gas Turbines & Power* 122(1): 135–40.

Smithsonian Institute. 2005. Collecting the History of Fuel Cells. *Smithsonian National Museum of American History.* americanhistory.si.edu/fuelcells/index.htm, accessed August 14, 2005.

Staka, C. 2002. Local Energy Policy and Smart Growth. *Local Environment* 7(4): 453–58.

Stone, D. 1988. *Policy Paradox and Political Reason.* New York: Harper Collins.

Tanrioven, M and Alam, M. 2005. Reliability Modeling and Assessment of Grid-Connected PEM Fuel Cell Power Plants. *Journal of Power Sources* 142: 264–78.

Tse, L. 2005. Molten Carbonate Fuel Cell. *Vision Engineer.Com*, www.visionengineer.com/env/mc.php, accessed August 15, 2005.

United Nations Environmental Programme (UNEP). 2003. Indian Solar Loan Programme. www.uneptie.org/energy/act/fin/india/, accessed November 27, 2005.

Vaitheeswaran, V. 2004. Unraveling the Great Hydrogen Hoax. *Nieman Reports* 58(2): 14–17.

Van den Broek, H. 1993. Research, Development, and Demonstration of Phosphoric Acid Fuel Cell Systems, in Blomen, L. and Mugerwa, M., eds. *Fuel Cell Systems.* New York: Plenum Press, pp. 245–70.

Vignola, F., Hocken, J., and Grace, G. 2000. *PV in Schools.* Eugene, OR: Oregon Million Solar Roofs Coalition.

Wald, M. 2004. Report Questions Bush Plan for Hydrogen-Fueled Cars. *New York Times*, February 6, A20.

White House. 2003. Fact Sheet: Hydrogen Fuel—A Clean and Secure Energy Future. www.whitehouse.gov/news/releases/2003/02/20030206-2.html, accessed November 27, 2005.

Wustenhagen, R. 2003. Sustainability and Competitiveness in the Renewable Energy Sector. *Green Management International* 44(Winter): 105–15.

Zuttel, A. 2003. Materials for Hydrogen Storage," *Materials Today.* 6(9): 24–33.

WEB SITE

chemelab.ucsd.edu/ fuelcell/soxide.htm, accessed August 14, 2005.

www.energy.ca.gov/afvs/clean_diesel.html, accessed August 14, 2005.

www.fctec.com.

www.fossil.energy.gov/programs/powersystems/fuelcells/fuelcells_moltencarb.html, accessed August 14, 2005.

www.hydrogencomponents.com/hythane.html, accessed August 14, 2005.

www.visionengineer.com/env/mc.shtml, accessed August 14, 2005.

Chapter Nine

Historical Precedents: Alternative Energy/Fuels and Legitimacy Issues

INTRODUCTION

A treatment of alternative energy as a concept would not be complete if due regard were not offered to historical precedent in the field of alternative energy. Considering the various forms of alternative discussed thus far, it is evident that each has experienced a period of intense research activity—many are subject to continued research and development. In the early stages of development, many energy alternatives may have seemed a bit far-fetched but later turned out to be valuable contributions to the energy paradigm. Conversely, other alternatives may have seemed quite practical from their origins, but are increasingly subject to intense criticism and scrutiny.

In the case of each form of energy generation or fuel, the technical, economic, and political feasibility dimensions were discussed. It is the latter issue that is crucial to the legitimacy of an energy source or fuel development plan. Unlike technical and economic feasibility, both of which claim a basis in knowledge and rationality, political feasibility is a function of values. The birthright of many forms of alternative energy cannot be claimed by the private sector alone.

Chapter 9 presents a model of public policy built around political and social values undergirding public policy. The chapter emphasizes a very crucial point: namely, variations in risk assessment in a pluralistic political and social system. Risk assessment plays a significant role in shaping the level of legitimacy of public policies. As Mary Douglas and Aaron Wildavsky (1983) posit, the level of acceptable risk in the contemporary society is quite low and getting lower. In other words, alternative energy must be technically and economically feasible as well as being nearly bereft of risk to humans and, in an environmentally conscious nation and world, to the natural environment in a very inclusive sense.

This chapter will:

- discuss the risk and culture hypothesis;
- relate the aforementioned hypothesis to the energy policy area; and
- illustrate evolving notions of risk in relation to energy policy through case studies of hydropower and nuclear energy.

RISK AND CULTURE: ALTERNATIVE ENERGY AND HIDDEN COSTS

At this point, the book has covered quite a bit of material and yet the question arises: What is alternative energy? An "alternative" to what? Is it renewable energy? If so, what standards make it renewable or nonrenewable? In my experience, when these questions arise, two or three things are likely to occur in the form of verbal or nonverbal response from fellow conversants:

- Reflexive defensiveness arises, as if to say, "Well, exactly who are you [your name here], and what's with this questioning of accepted 'fact?'"
- Deer-in-the-headlights looks arise when the uninformed or undecided face these questions.
- Condescension arises when the conversant thinks that the questioner is uninformed. If questions are asked in apparent innocence, then this response often follows initial reflexive defensiveness.

So, is this just sarcasm or do things really happen like this? It's not sarcasm. Yes, things like this really occur and not just in alternative policy venues. The question arises as to why this occurs and why this might actually be important in understanding the evolving definition of alternative energy and renewable energy, albeit the latter term may have achieved greater solidity than the former simply due to obvious outcomes. It is, of course, important to overcome this dilemma if sustainable communities are to be developed in an equitable and collaborative manner. Communication and basic acceptance of an overarching policy narrative must occur or innovation never moves from basic premises toward effective conclusions about public policy and the nature of the sustainable community. Understanding the risk of doing something or nothing at all must also be a part of the policy narrative.

In their now-famous account, *Risk and Culture: An Essay on the Selection of Technological and Environmental Dangers* (1983), Mary Douglas and Aaron Wildavsky come to terms with public policy direction through a study of risk assessment. Ultimately, they argue, our selection of public policy, particularly in areas such as environment, crime, and public health, are a function of what is feared most by the collective citizenry.

In life, risks abound. Many risks are judged, consciously or not, to be of little importance. Risks may be brushed aside, despite the fact that substantial evidence points to a high level of risk. Times change and perceived or real levels of risk are adjusted. For example, smoking cigarettes was once thought to be good for one's health. Since the 1960s, public policy efforts have focused on informing smokers of the potential dangers. In essence, as Douglas and Wildavsky (1983) would say, "Risk should be seen as a joint product of *knowledge* about the future and *consent* about the most desired prospects."

In school, one suspects that students are often taught "the facts" about things of various sorts. There is a sense that the facts are immutable truths about the world and how it works. As Douglas and Wildavsky (1983) argue, however, knowledge can be contested, thus making it very difficult to establish agreed-on facts shaping risk assessment. A great example of this dilemma arises in the global warming debate. Scientists, politicians, interest groups, and interested citizens all have quite disparate views on whether global warming is occurring and whether the use of fossil fuels is negatively impacting the environment. The basis of knowledge is murky and facts are disputed.

In other cases, information is fairly certain and is not contested in the risk assessment process. A good example comes from the economics of energy policy arena. Few would disagree that the basic economic principles of supply and demand will impact energy prices. While the amount of available fossil energy is contested, it is generally accepted that, ceteris paribus, as demand increases in relation to supply, fuel prices will increase and likely impact economic outcomes for individuals, nations, and the world as a whole.

The second dimension in the analysis of risk assessment is related to political *consent*. Although likely to exist on a continuum, consent ranges from *complete* to *contested*. An example of complete consent might be found in the area of energy economics—specifically, in energy policies for the indigent. Very few individuals would openly deny access to energy for poor citizens and most would accept the idea that the poor should either pay reduced rates or nothing at all for access to electrical energy.

According to Douglas and Wildavsky (1983: 5), when information is certain and not contested, then problems associated with a risk are thought to be *technical* in nature. The "solution," therefore is viewed as a matter of "calculation." In the example from energy policy in the previous paragraph, the risk associated by rising global demand for fossil fuel—particularly petroleum—can be studied through economic modeling. Currently, economists have found that the potentially negative economic impacts of rising oil prices have been offset for nations, such as the United States, by other macroeconomic behaviors. It is also not generally contested that increasing energy supply will reduce energy prices and reduce risks to economic growth for individuals and society as a whole, although exactly how to go about increasing energy supply is highly contested in other ways (see figure 9.1).

When knowledge is certain and consent is contested, there frequently exists significant disagreement in risk assessment. According to Douglas and

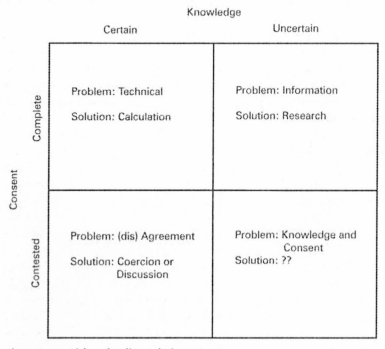

Figure 9.1.   Risk and Policy Solutions
Source: Douglas and Wildavsky (1983: 5). Art reproduced by author.

Wildavsky, the definition of risk and policy solutions to limit defined risk becomes a function of political coercion or discussion. Thinking back to an earlier discussion of Lowi's policy typology in chapter 2, public policy of this type is often subject to the influence of interest group politics, as consent to risk assessment develops. In Lowi's typology, policies most likely to be a function of interest group politics are either redistributive or regulatory. As will be discussed later in the chapter, advocates for nuclear energy would argue that the knowledge undergirding nuclear energy is certain but that consent is contested; from the perspective of a nuclear energy proponent, the primary reason why nuclear energy has not been allowed to develop further in the United States is related to interest group politics.

When knowledge associated with risk assessment is uncertain and consent is complete, then the primary problem facing risk assessment is a lack of information to deal with limiting or ameliorating the impact of a risk. The solution, therefore, is to pursue or encourage research associated with a particular risk. In the case of energy policy, significant resources are being put toward research and development of alternative energy solutions to a perceived risk of energy supplies not being capable of meeting demand and the resulting risk (or certitude) of facing higher energy costs and lower economic growth.

Government research funds for energy development are by no means a recent development. Research funding for fossil fuel research has been going on for nearly a century, with the establishment of the Coal Research Center by the Bureau of Mines (BOM) in 1910. In 1918, the BOM, a subunit within the U.S. Department of Interior, established the Petroleum Experiment Station in Bartlesville, Oklahoma (see National Energy Technology Laboratory 2005 [www.netl.doe.gov/, accessed August 19, 2005]). With increased emphasis on accessing petroleum and natural gas reserves in Alaska, the National Energy Technology Laboratory opened its Arctic energy office in 2001. In all instances, the focus has been on continued research and development of fossil fuel, effectively reinforcing its role in the energy paradigm for some time to come.

According to Douglas and Wildavsky's (1983) model, when knowledge is uncertain and consent is contested, there is no easily predictable solution to the "problem." In fact, the problem is itself contested in definitional form. For some individuals and groups, it might not be a problem at all, while other individuals and groups find that the problem is tantamount and must be addressed. It is in this fourth quadrant of the model of risk assessment where Douglas and Wildavsky are most interested in describing and it is an area that becomes of critical importance to any individual studying or thinking about energy policy of the future. It is a region, a quadrant, in which nearly all energy policy first emerges, a region where science is unclear and the nature of consent or contestability has yet to be determined.

When it comes to risk assessment, the "center" is complacent (Douglas and Wildavsky 1983). For the most part, most individuals are not deeply interested in the scientific foundations of public policy, nor do they have deepseated ideologically driven value structures that lead either to the acceptance or outright rejection of policy (or lack thereof) in relation to a level of risk. Although Douglas and Wildavsky (1983) do not directly emphasize this point, it is clear from the temper of their writings that the center is driven primarily by the maintenance of established political, social, and economic patterns. The individuals and groups along the fringes of society and politics, the "border" (Douglas and Wildavsky 1983), tend to be more interested in defining risk either through the lens of new scientific theory or in relation to a set of core values divergent from the majority or both. In any event, the border tends to be more risk adverse when it comes to status quo—change will reduce perceived risks.

Douglas and Wildavsky (1983) conclude that the border is what tends to be more active in shaping the policy process, from agenda setting through formulation, implementation, and evaluation, while the center tends to be largely disinterested in risk assessment. The border seeks to gain the attention and interest of the center not through direct appeals to a deep-seated knowledge base or deeply held values; rather, the border attempts to influence the center by appeals to an interest in maintaining economic, social, and political *conditions and benefits*, which—it is advocated—can only be achieved through margin or possibly radical change in *methods* of existence.

At times, things are made easier for the border through societal (perhaps global) shifts in basic values (see Inglehart 1990), which lead to shifts in consent and alter the nature of risk assessment and public policy. Increased levels of education may shape individuals' ability to comprehend risk assessment, conceptually identify multiple perspectives of risk, and proactively critique and accept/reject its validity and generalizability. In other instances, events occur that provide individuals and groups along the border opportunities to employ phenomenology to represent their claims to knowledge or a core set of political, social, or economic values. For example, an exceptionally warm summer in a city is used as an opportunity for certain actors along the border to promote green energy and a reduction in greenhouse gas emissions. Ultimately, the center remains complacent, concerned primarily with its political, social, and economic self-interest, the maintenance and continued improvement of a lifestyle.

Evidence for the aforementioned views of risk assessment can be found in a case study analysis of hydroelectric dams and nuclear energy policy. One will find that the legitimacy of or at least weighted value of knowledge and levels of consent change over time. Times change; the policy environment surrounding alternative energy has and continues to evolve. Claims to knowledge and levels of consent have and continue to shape our risk assessments associated with energy policy.

HYDROELECTRIC DAMS

Hydropower or water power is one of the oldest sources of energy used in human civilizations. Water power was primarily used to operate mechanical equipment used in processing grain, which remained an important use of water power for millennia. While electricity-related research had occurred for centuries, it was not until the application breakthroughs of the 1870s that commercial electricity became practical.[1]

The first hydroelectric dam in the United States began operations on September 30, 1882. The dam was built over the Fox River in Appleton, Wisconsin (EIA 2005 [http://www.eia.doe.gov, accessed March 22, 2006]). Other dam projects quickly followed in the late nineteenth century, most of them either privately owned or controlled by nascent urban public utility boards. The image of hydropower during the late nineteenth and early twentieth centuries was that of freedom and economic growth. Electricity was synonymous with urban development. Electric mass transit systems appeared. Electric light meant that workers could operate machines during nighttime hours. Previously, homes were lit by expensive candles and oil, but electricity reduced the price for illumination. While electricity was not generated solely by hydropower, it remained an important source of supply (see Jonnes 2004).

Electric energy generation was controlled by local utilities and private electric companies who worked with local utility boards. The energy market was seen as a combined local government and private corporate function. The political feasibility of federal government involvement in energy generation expanded with the passage of the Tennessee Valley Authority Act of 1933 (16 U.S.C. 12; see also Abrams 1937) and the Bonneville Project Act of 1937 (16 U.S.C. 832). Signed in 1935, Franklin D. Roosevelt's Executive Order 7037 established the Rural Electrification Administration, forging a strong federal role in the management of electric power generation and distribution (Campbell 2000). Despite movements toward deregulation in the 1990s, the impact of the 1930s policy shift still impacts energy policy and is perhaps most visible in terms of hydropower projects.

Hydropower provides approximately 45 percent of all forms of renewable energy used in the United States (Sale 2005)—less than 3 percent of all energy consumed. Hydropower is slightly greater than wood in terms of its share renewable energy production, the latter remaining a significant part of the renewable energy consumed. By contrast, solar, wind, and geothermal sources make up less than 1 percent of the total energy consumption in the United States (Sale 2005), which have been a centerpiece of the discussion of next generation renewable green energy generation. Hydropower is often contested as being a desirable approach to a green energy future.

The economic, technical, and political feasibility of hydroelectric and flood control dam projects has been questioned for quite some time (Baker 1988; Howarth 1960; *National Wildlife* 1984) and alternative plans for balancing human needs with ecological realities (see Marts and Sewell 1960). The Sierra Club and other environmental interest groups historically have been quite active in opposing the installation of dams. Early on, environmentalists and ecologists questioned the hidden costs of dam installation in general, and hydroelectric dams in particular. All dams act as natural barriers to stream flow and migration of sediment and wildlife in streams and along stream banks. Hydroelectric dams are particularly damaging to salmon seeking to move upstream to lay eggs in native spawning grounds. A system of fish ladders have been designed to encourage fish migration upstream by bypassing dams. Dam breaching is another method of improving salmon movement through streams and rivers (Tatro 1999).

State-level ecology movements in Oregon in 1980s and 1990s, for instance, effectively culminated in statewide efforts in the 1990s to bring attention to the issue of technical feasibility issues as well as the dynamics of public consent to a standing policy (see Nicholas 1997). The state policy offers clear evidence that grassroots policy efforts have a significant role in offering legitimacy to scientific/technical feasibility as well as offering or withholding political/social consent. The technical feasibility and consent may not exist entirely on separate dimensions as the Douglas and Wildavsky model would indicate—in essence, there is an interaction effect in ultimately assessing policy risk and benefit. Furthermore, state and local grassroots efforts in shaping the risk and benefit dialogue are not provincial. A bottom-up policy dimension emerged from Oregon's evolving consideration of its long association with the hydroelectric paradigm. Federal-state partnerships have emerged since the National Marine Fisheries listed coho salmon as an endangered species (see *American Forests* 1999). As Steel et al. (1999) found in their study of salmon recovery in relation to the hydroelectric policy paradigm, policy implementation surrounding the issue of policy balance is directly tied to knowledge and values, and energy policy is by no means exogenous to policy developments in related arenas.

The Pacific Salmon Treaty Agreement, Endangered Species listings of salmon and steelhead, and the Pacific Coast Salmon Recovery Fund (see 16 U.S.C 56A §3645 and P.L. 106-113) test the political feasibility of hydropower and effectively expose some of the heretofore less visible economic costs of hydropower electricity generation. By redefining the issue of technical feasibility to include the ability of hydropower to eschew large-scale damage to the environment, the act effectively sends hydropower experts back to the drawing board to meet new standards of technical feasibility—standards that will meet the political and social values of a new policy era seeking a reduced level of environmental risk. At the very least, the salmon issue and subsequent policy adjustments related to hydropower generation have "moved" hydropower from the upper-left quadrant in Douglas and Wildavksy's model to the lower-left quadrant. It is possible that with further erosion of the knowledge base related to hydropower and its impacts, the technical feasibility of the electricity generation model will be more fully questioned, moving risk assessment into an unknown region of the lower-right quadrant—that is, the quadrant of contested definitions and acceptance related to the nature or existence of a problem and a potential solution.

NUCLEAR ENERGY

Nuclear energy is largely a twentieth-century phenomenon. The theoretical work of Albert Einstein and critical demonstration projects of Enrico Fermi and Leo Szilard in 1940 showed that although very small, atoms contain a tremendous energy potential, particularly when broken apart, a process known as fission. Uranium-235 was particularly interesting to physicists studying nuclear energy because when the element encounters an additional neutron, it becomes unstable and breaks apart or "decays." In the process of decay, uranium releases large bursts of thermal and radioactive energy. The decay process becomes a chain reaction, as neutrons released from one fission event impact other uranium atoms.

The thermal energy released from fission reactions is quite significant and can ultimately produce high-pressure steam ($\sim$315° C) that can be used to turn liquid water into steam. Steam can be used to operate turbines for the generation of alternating current. In most U.S. reactors, the water in piping that comes in contact with nuclear material is in a closed loop (frequently, these systems are pressured water reactors) and thus more likely to be contained in the event of a reactor failure. The heat from the steam produced in the closed loop is transferred via heat exchanger to another system, usually water-based, that creates the steam needed to operate the turbine system (http://hyperphysics.phy-astr.gsu.edu/hbase/nucene/reactor.html#c3, accessed August 29, 2005).

Nuclear energy proponents argue that nuclear energy is relatively safe. Nuclear energy plants cannot blow up like nuclear bombs but could conceivably produce small explosions and fires if the energy plant system overheats or in some other way malfunctions in operation. Through peer-reviewed publication, demonstration projects, and everyday evidence from commercial plant operations, nuclear engineers and nuclear physicists lay claim to a well-grounded scientific knowledge base regarding the operations of nuclear energy plants in commercial energy production settings.

Perhaps the strongest piece of evidence in the quivers of nuclear energy advocates relates to the empirical reality that nuclear energy safely provides 20 percent of the electrical energy needs in the United States and produces the vast majority of electrical energy needs in many European nations—for instance, 77 percent of the electrical energy needs in France are provided by nuclear energy (EIA 2005).

There are, however, many aspects of nuclear energy technology that remain somewhat elusive. Perhaps the most commonly discussed issue is how to process and manage the radioactive waste products of nuclear energy (see Dolan and Scariano 1993). In 2000, the U.S. government itself shipped approximately 2,467 metric tons (approximately 2,719 short tons) of spent nuclear fuel (SNF) (DOE 2002: 2) to nuclear waste repositories in the United States. The vast majority of this SNF is stored at the U.S. Department of Energy (DOE) site at Hanford, Washington.

The science behind nuclear waste management and disposal is of significant concern because dangerous levels of radioactivity are present in commercial nuclear energy waste products for, quite literally, tens of thousands of years. As mentioned previously, the Hanford site in Washington State has until recently been accepting the vast bulk of nuclear waste from government sources and some medical- and research laboratory-related radioactive wastes, while simultaneously operating one of the nation's largest nuclear waste decontamination operations on the facility lands—a remnant of several decades of nuclear weapons facility operations during World War II and the subsequent Cold War.

In November 2004, the voters of Washington State passed a ballot initiative with a 69 percent majority in favor of banning nuclear waste imports into the Hanford site. The U.S. Department of Justice challenged the ballot initiative in federal court (*U.S. v. Hoffman*), arguing that "federal law governing nuclear waste shipments should prevail over state law and asserts 'sovereign immunity' over the waste generated by weapons production" (Steele 2004: B1). The federal court ordered a temporary injunction against the ballot initiative's implementation, requesting a decision from the state Supreme Court of Washington State with regard to the intent of the initiative in terms of challenging federal authority over Hanford site management. In July 2005, the Washington State Supreme Court rejected the federal government's argument. The Court concluded that the ballot initiative "was drafted to prevent the addition of new radioactive and hazardous waste to the Hanford nuclear reservation until the cleanup of existing contamination is complete" (Cooper 2005).

Beyond the issue of government-created nuclear waste is, of course, a much larger concern over the commercial SNF that is produced by nuclear power stations across the nation. Since the Three Mile Island incident in 1979 in which a reactor core overheated, no new nuclear power facilities have come on line for commercial power generation. Critics of nuclear energy argue that these waste materials have the potential added cost of destroying the environment and harming animal and plant life well into the foreseeable future, quite possibly beyond the existence of human beings as a species. The Government Accounting Office (GAO) estimates that as of 2003, there are over 50,000 tons of spent commercial nuclear fuel "stored at 72 sites at or near nuclear power plants in 33 states" (GAO 2003: 1).

U.S. government responsibility over the regulation and management of nuclear fissile materials can be traced back to the Atomic Energy Act of 1954. In 1982, Congress passed and President Reagan signed into law, the Nuclear Waste Policy Act (NWPA). NWPA directed the DOE to conduct studies of appropriate locations for the storage of commercial and government-generated nuclear waste in a permanent repository. In 1987, "Congress amended the act . . . and required DOE to only consider Yucca Mountain, Nevada, as potential site for a repository. In 2002, the president recommended to the Congress and the Congress approved, Yucca Mountain as a suitable site for the development of a permanent high level waste repository" (GAO 2003: 5).

In 1995, the DOE, U.S. Navy, and the state of Idaho entered into a precedent-setting agreement on the movement of nuclear waste within the state. In essence, the DOE agreed to limit shipments into waste repositories within the state and in return promised to remove nuclear waste from the state sites and transport it to safely transport it from the Idaho National Engineering and Environmental Laboratory (INEL) to a permanent repository. The timeline established in the document indicates that all nuclear waste will be removed from INEL by 2035. The agreement essentially established a timeline for the removal and transportation to the permanent waste repository. The agreement pushes forward the process of permanent nuclear waste disposal and reduces the likelihood that waste will continue to be transported and stored at temporary repositories.

Despite the growing sense that there needs to be a permanent repository established, there remains significant controversy over the use of Yucca Mountain as a nuclear waste depository. Environmental interest groups have historically been opposed to nuclear energy because of the potential of harm to humans and to the environment in energy production and waste storage; concerns that have been fueled by evidence from nuclear power accidents at Three Mile Island in 1979 and at Chernobyl in the former Soviet Union in 1986. Nuclear waste storage was further hampered by a general toxic waste-dumping scandal that occurred in the 1970s that produced images of rusting harmful toxic waste containers left in shopping center parking lots and in pristine wooded areas. In short, images can prove to be quite powerful and long lived in the minds of many individuals.

Nuclear waste repositories exist at the aforementioned seventy-two storage sites. The Nuclear Regulatory Commission (NRC) has approved dry pellets stored in steel containers within cement structures or in shallow cooling ponds constructed of several feet of steel and cement designed to prevent contamination. Concerns remain over the ability to manage nuclear waste spread over many different sites, but there are equal concerns about the transportation to and safe storage of nuclear waste materials at a repository site such as Yucca Mountain, Nevada.

In 2003, the GAO conducted a study to determine the level of safety associated with transportation of nuclear waste via truck or rail to Yucca Mountain. In the past, major concerns had focused on nuclear waste spills produced by transportation accidents. The GAO report recounted these analyses, finding that vehicle accidents impacting the transportation of fissile waste materials would occur in the range of four to seven times per one hundred thousand rail or truck accidents, which were considered minimal likelihood in statistical terms alone. Another major concern that emerged in the report was related to terrorist attacks on the transportation of commercial nuclear waste to Yucca Mountain. The GAO study concluded that the cement and steel casks in which the nuclear materials were transported would most likely survive a terrorist attack and that only a series of highly improbable events would result in the release of radioactive contaminants. Citing DOE research evidence, the GAO found that confidence in the very high level of safety associated with transportation of nuclear waste to Yucca Mountain has continued to grow due to increased amounts of data and growing sophistication of simulation programs to study potential weaknesses in the transportation and disposal model (GAO 2003: 11–12).

A secondary method of rejecting scientifically-defined risk in terms of nuclear waste disposal and power plant decommissioning could involve appeals to the Environmental Protection Agency (EPA) to develop secondary risk assessments and to prevent the NRC, which is more closely tied to energy producers than to environmental interests, from regulating NRC waste and plant decommissioning activities. At least for the moment, the policy "door" remains closed to interests pursuing the aforementioned strategy because of a 2002 Memorandum of Understanding in which the EPA expressly denies itself the ability to regulate the activities of the NRC in relation to nuclear material transportation and commercial nuclear energy plant-related environmental measures, most specifically those associated with plant decommissioning (Whitman and Meserve 2002).

Risk assessment, however, remains the central focus of concern with regard to nuclear waste, as detailed in the recently NRC report *Risk and Decisions: About the Disposition of Transuranic and High-Level Radioactive Nuclear Waste* (2005). The NRC eschews offering specific recommendations, focusing instead on the ways in which risk is defined in relation to nuclear waste. The NRC has reviewed a number of previous studies conducted by government and nongovernmental organizations regarding the level of risk associated with the transportation and storage of nuclear waste. The council paid particular heed to the politically charged atmosphere surrounding the definition of risk, identifying the level of angst expressed through ballot initiatives and litigation, extant in a multitude of state and local communities: in short, there is very little trust in the process of dealing with nuclear waste.

Views of nuclear waste management vary significantly across policy stakeholder groups. Native American tribes, so-called downwinder communities— communities that were impacted by past nuclear radioactive contamination— and environmental groups demonstrate a high level of distrust of scientifically defined risk assessment produced by governmental agencies and by nuclear industry policy analysts. One thing that can be agreed on is the need to remove nuclear contamination and waste; but the big questions relate to final repository location and transportation: in brief, it is the "not-in-my-backyard," or NIMBY effect, as well as historically demonstrated and now legally recognized damaging impacts of nuclear energy development and waste storage (Washington Department of Public Health 2004).

The economics of nuclear energy are quite revealing and offer a different view of how risk is assessed. Nuclear energy as a source of electricity generation represents a significant portion of the net power generation portfolio. Currently, 104 nuclear power plants produce approximately 20 percent (as of 2003, 763.7 billion kWh) of the commercial electrical energy in the United States (U.S. Census Bureau 2003–2004: 587), making nuclear energy the second-largest energy production sector behind coal-fired electricity plants (EIA 2005 [www.eia.doe.gov/emeu/cabs/usa.html, accessed August 29, 2005]).

The ratio of operating capacity to generation capacity has increased tremendously over recent years. But energy production comes at a price in terms of nuclear material used. The process of mining, enriching, and fabricating nuclear fuel is estimated at $1,633 (2006 dollars) per kilogram, which would yield 315,000 kWh. In terms of uranium concentrate purchases alone, current nuclear energy costs are around $0.0052 per kWh (www.uic.com.au/nip08.htm, accessed September 26, 2006).

Capital costs for the construction of nuclear power facilities are not presently being factored into the cost of nuclear-produced electricity in the United States because new plants are not coming on line and older plants are largely bought and paid for at this point. Nevertheless, if new plants were to be built, overnight capital costs would largely be a function of the number of plants being built, the size of the plant, time to completion, and the expected lifespan of the facility. Plant costs for a next generation 1,500 kWh plant, for instance, are between 6.2 and 3.4 cents per kWh produced over the plant lifespan. Capital costs per kWh for power generation would decline significantly if a commitment to large-scale plant construction was undertaken nationwide.

It is important to note that plant cost estimates are changing because of technological developments, such as the pebble bed modular reactor (PBMR) design. PBMR is an energy option that has been explored globally. The system relies on nuclear fuel "pebbles" composed of uranium oxide and graphite, the latter intended to help manage the fission process and reduce the chance of overheating (although overheating associated with graphite and graphite fires was part of the problem in the Chernobyl nuclear plant accident in 1986). The pro-PBMR analysis at University of Michigan claims that the cost of these advanced reactor units is much smaller than a conventional nuclear plant, requiring less down time for fuel replacement and fewer operating personnel. These two factors alone, it is claimed, will significantly reduce operations and maintenance costs for electricity generation on a per kWh basis. The plant facilities are modular and require a smaller land "footprint," which will also reduce capital cost expenditures. Anti-PBMR studies, however, claim that the plant design is unsafe and could result in the release of radioactivity into the atmosphere. Critics find that the pebbles have been shown to have at least one defect per pebble, which could result in uncontrollable fission reactions and thus overheating.

The debate over the quality of pebble materials, the relative safety of nuclear power generation, and the "true cost" per kWh of nuclear energy remains mired in politically charged analysis. The issue of managing nuclear waste, however, is an area where pro-nuclear forces have, in recent years, provided new analysis that might change the nature of debate. European innovations have resulted in significant reductions in SNF through reprocessing and reuse of radioactive material. Reprocessing recovers usable plutonium and uranium; the basis of a product known as mixed oxide fuel (MOX). MOX does not require enrichment because it has already been processed as a fuel source, which reduces production costs of fuel. Additionally, resulting SNF, which cannot be used as a source of MOX, is estimated at only 35 percent of the "volume, mass, and cost of disposal" when compared to traditional SNF, which is not subject to the MOX recovery process. In Germany and other European nations, nuclear reactors are designed to include MOX as part of the energy generation process and, it is claimed, this reduces the cost of energy generation (www.uic.com.au/nip08.htm, accessed September 26, 2006]).

The Economic Future of Nuclear Power (Tolley and Jones 2004), a report produced by a team of economists and other public policy analysts at the Harris School of Public Policy, is a recent and widely studied analysis of nuclear energy at the dawn of the twenty-first century. The report was delivered to the U.S. House of Representatives and has been subjected to peer review as well as government administrator review and likely serves as an important steppingstone in DOE analysis produced in 2005 and reported in the next section.

The University of Chicago study focused on the levelized cost of electricity for current and next-generation nuclear power plants when compared with projected costs for other major sources of electrical energy generation likely to be used in the new century. The study found that initial analysis of the levelized cost of electricity of nuclear energy is somewhat higher than electricity from "conventional" energy sources (i.e., coal, gas-fired, and oil-fired plants). A new 1,000 MW nuclear power plant is estimated to cost between approximately $550 million and $600 million dollars in terms of overnight capital costs (for seven-year construction time), compared to coal and gas generation plants, which are estimated to have overnight capital costs of between $275 million and $375 million. On a per MWh basis, the range for nuclear is between $47 to $62, while the range for coal and gas-fired plant capital construction costs is $33 to $45. The University of Chicago report, however, demonstrated that capital costs are expected to decline significantly *if* there is a commitment to build multiple nuclear power plant units.

Two additional factors make nuclear energy economically attractive during the twenty-first century. First, although the costs of coal are expected to remain fairly stable, the costs of natural gas may increase significantly by mid-century, possibly reducing the current cost per kW advantage that gas has over nuclear energy in terms of cost of electricity. In terms of nuclear fuel costs, uranium prices have declined significantly over the last twenty-five years are expected to remain relatively low over the current century. Second, greenhouse gas emissions policy may drive up the levelized cost of electricity generated using coal or gas, whereas nuclear energy as a low emission method of generating electricity will become more economically and environmentally attractive in terms of emissions factors. As discussed previously, states have established emissions standards and reduction incentives (see EPA 2000) and the federal government has provided strict guidelines for emissions as well (e.g., Clean Air Mercury Rule, Clean Air Act, EPA Tier II standards, Clean Air Act Emissions Reduction Amendments "Clear Skies Act of 2003"—42 U.S.C. 7651, et seq.). Although the United States is not a signatory nation for Kyoto, its emissions policies have been impacted by international standards. In the long term, this will likely make nuclear a more politically and economically viable energy solution. The cost of nuclear energy in terms of capital expenses and operations and maintenance costs may be offset by federal and state emissions standards limiting conventional fossil energy power generation combined with limited renewable green energy generating capacity. If the federal government provides significant tax credit incentives to nuclear energy facilities constructed in the future, then the cost of electricity generated using nuclear power will become more competitive with current conventional power plants.

THE REBIRTH OF NUCLEAR ENERGY?

A simultaneous development has been a growing interest in the expansion of nuclear energy as part of the national energy portfolio. The recently published Nuclear Energy Task Force report, *Moving Forward with Nuclear Power: Issues and Key Factors* (2005), builds on themes similar to the University of Chicago study. The report seeks to streamline the process of permitting for new nuclear power facilities. The study outlines methods of achieving this goal, calling on the DOE to preapprove nuclear facility designs to reduce the time commitment necessary for each facility application and approval process. The process of design and site approval is costly and the litigation that is likely to appear during the operations approval phase is an obstacle to firms that are interested in building and operating nuclear power facilities.

Reducing the uncertainty of plant design and operations through preapproval of plant designs and facility operations plans was found by NETF members to be a significant step forward in developing nuclear energy plants in the future. Critics would likely charge that the NETF recommendations are simply methods of reducing public response time to the development of nuclear power plant facilities. The report does spend minimal effort to promote the principles of sustainable communities with regard to public visions of acceptable energy development. In other words, the role of consent is firmly embraced in the study, rather than a narrow focus on technical feasibility. The NETF study is a substantial move forward in the further development of nuclear energy as a macro top-down energy policy solution to meet short- and long-term needs. Political outcomes over the next decade will likely determine if macro top-down solutions, such as the further development of nuclear power, will become policy realities or not. Given the need to remain competitive in an international market-driven economy, I anticipate that a resurgence of nuclear energy in some form will occur, but remain part of a larger mixed-source energy portfolio. The NETF study represents an attempt at a technical solution to substantially reduce potential barriers.

CHAPTER SUMMARY

Hydropower and nuclear energy are two excellent examples of the politics and science of energy policy. In both cases, the energy sources were initially seen as built on solid scientific knowledge and were largely uncontested in terms of political and social consent. In the 1970s, the energy policy paradigm was shaken significantly as science and political and social consent for these two energy sources began to change. The science of hydropower was questioned as ecological evidence increasingly documented the impact of hydropower dams on the natural environment. Political and social consent to the risks associated with hydropower was significantly eroded. Nuclear energy faced similar changes in risk assessment, particularly following the Three Mile Island incident.

Despite eroding consent, the reality is that a significant portion of the national energy portfolio is met by hydropower and nuclear power. Risk assessment aside, the price of electrical energy is a third dimension. Consumer choices in the energy market tend to be driven to a great extent by often inelastic demand, rather than by thoughtful choices about energy production sources. In light of this fact, it is likely that the risk assessment of hydropower and nuclear energy will face increased scrutiny as consumers face rising electrical energy prices over the next century.

NOTE

1. In 1879, Thomas Edison unveiled his light bulb and several electrical machine applications through what is now called General Electric. Design and implementation of urban electrification began to occur and demand for electricity grew. Supply of

electricity, however, was and remains a perennial issue in the United States. Using fossil energy to power turbines for electricity generation was one method of meeting demand, particularly since the post–Civil War era witnessed tremendous crude petroleum discoveries in eastern Pennsylvania and other states. For a short period of time, supply was outstripping demand. Perhaps unbelievably in the current era of price increases, the price of oil fell precipitously from approximately $70 per barrel (in 2004 U.S. dollars in the late 1860s to approximately $15 per barrel in the mid-1870s (in 2004 U.S. dollars) (see WTRG Economics 2005 [www.wtrg.com/prices.htm, accessed August 20, 2005]). Nevertheless, petroleum was not readily accessible to all due to transportation costs; therefore, other sources of electricity generation were developed, including hydroelectric dams. While technical feasibility issues had been overcome through decades of electromechanical research, economic feasibility of electricity was a function of the price of energy needed to operate generation facilities; hydroelectric power lowered energy costs and made generation economically feasible in many locations.

WORKS CITED

Abrams, E. 1937. Your Stock in TVA. *Saturday Evening Post* 210(16): 27, 77–82.
American Forests. 1999. Salmon Efforts Jumping. *American Forests* 105(1): 11–12.
Baker, J. 1988. Letting the Rivers Flow. *Sierra* 73(4): 21–24.
Campbell, D. 2000. When the Lights Came On. *Rural Cooperatives* 67(4): 6–9.
Cooper, B. 2005. *State Supreme Court Upholds Key Provisions of Hanford Cleanup Initiative 297.* www.hoanw.org/PDF/7_28_5.pdf, accessed August 29, 2005.
Dolan, E. and Scariano, M. 1993. *Nuclear Waste: The 10,000 Year Challenge.* New York: Franklin Watts, Inc.
Douglas, Mary and Wildavsky, Aaron. 1983. *Risk and Culture: An Essay on the Selection of Technological and Environmental Dangers.* Berkeley: University of California Press.
Energy Information Agency (EIA). 2005. *Country Analysis Briefs: France.* www.eia.doe.gov/emeu/cabs/France/Electricity.html, accessed November 27, 2005.
Environmental Protection Agency (EPA). 2000. *Federal and California Exhaust and Evaporative Emissions Standards for Light Duty Vehicles and Light Duty Trucks*, EPA420-B-00-001. Washington, D.C.: U.S. Environmental Protection Agency.
Government Accounting Office (GAO). 2003. *Report to the Chairman, Subcommittee on Energy and Air Quality, Committee on Energy and Commerce, U.S. House of Representatives: Spent Nuclear Fuel: Options Exist to Further Enhance Security.* GAO-03-0426. Washington, D.C.: Government Accounting Office.
Howarth, D. 1960. Giant in the Jungle. *Saturday Evening Post* 232(40): 26, 97–100.
Inglehart, Ronald. 1990. *Culture Shift in Advanced Industrial Democracies.* Princeton, NJ: Princeton University Press.
Jonnes, J. 2004. New York Unplugged, 1889. *New York Times*, August 13, A21.
Marts, M. and Sewell, W. 1960. The Conflict between Fish and Power Resources in the Pacific Northwest. *Annals of the Association of American Geographers* 50(1): 42–50.

National Energy Technology Laboratory. 2005. *History*. Washington, D.C.: U.S. Department of Energy. www.netl.doe.gov/, accessed August 19, 2005.

Nuclear Regulatory Commission (NRC). 2005. *Risk and Decisions: About the Disposition of Transuranic and High-Level Radioactive Nuclear Waste*. Washington, D.C.: National Academies Press

National Wildlife. 1984. Power vs. Wildlife, *National Wildlife* 22(3): 35.

Nicholas, J. 1997. *The Oregon Plan: Coastal Salmon Restoration Initiative*. Salem, OR: State of Oregon.

Nuclear Energy Task Force. 2005. *Moving Forward with Nuclear Power: Issues and Key Factors*. Washington, D.C.: Secretary of Energy Advisory Board, U.S. Department of Energy.

Sale, Michael J. 2005. *An Overview of Hydropower Resources in the U.S.* Oak Ridge National Laboratory. Renewable Energy Modeling Project, May 10.

Steel, B., Lovrich, N., and O'Toole, E. 1999. Public Perceptions and Preferences Concerning Pacific Salmon Recovery: An Oregon "Voluntary Grassroots" Perspective. *Social Science Journal* 36(3): 497–513.

Steele, K. 2004. Nuclear Waste Initiative Targeted. Challenge Contends Federal Laws Prevail over State Rules. *Spokesman Review* (Spokane, WA), December 2, B1.

Tatro, S. 1999. Dam Breaching. *Civil Engineering* 69(4): 50–55.

Tolley, G. and Jones, D. 2004. *The Economic Future of Nuclear Power*. Chicago.: Harris School of Public Policy, University of Chicago.

U.S. Census Bureau. 2004–2005. *Statistical Abstract of the United States 2004–2005*. Washington, D.C.: U.S. Census Bureau.

U.S. Department of Energy (DOE). 2002. *DOE Current Year Waste/Contaminated Media and SNF Inventories by Site (Sum-3)*. Washington, D.C., U.S. Department of Energy, Office of Environmental Management, Central Internet Database.

Washington Department of Public Health. 2004. *An Overview of Hanford and Radiation Health Effects*. www.doh.wa.gov/hanford/publications/overview/overview.html, accessed November 27, 2005.

Whitman, C. and Meserve, R. 2002. *Memorandum of Understanding between the Environmental Protection Agency and the Nuclear Regulatory Commission: Consultation and Finality on Decommissioning and Decontamination of Contaminated Sites*. Washington, D.C.: EPA and NRC.

WTRG Economics. 2005. *Oil Price History and Analysis*. London, AR. WTRG Economics. www.wtrg.com/prices.htm, accessed August 20, 2005.

WEB SITES

hyperphysics.phy-astr.gsu.edu/hbase/nucene/reactor.html#c3, accessed August 29, 2005.

www.uic.com.au/nip08.htm, accessed September 26, 2006.

Chapter Ten

Public Policy, Institutional Developments, and Policy Interests

INTRODUCTION

Public policy is often very complex. Circumstances may arise that lead adjustments in priorities and methods of achieving goals. Escalating energy prices, the global warming debate, and concerns about national political, social, and economic security all play roles in shaping priorities. Public policy is a function of citizen stakeholders and policymaker values. Individual values shape opinions and beliefs and ultimately impact policy preferences. Preferences and the expression of preferences in terms of advocacy or policy outcomes are shaped by another force observed in the form of institutional rules and roles. Rules are established to circumscribe the method by which we reach collective choices about collective benefit, as well as the range of possible choices that could emerge. With a certain degree of latitude, institutional roles assign positions and bases of influence to individuals within institutional settings. The interaction between varying perspectives of circumstance and solutions, individual values and interests, institutional rules and roles, the ever-evolving nature of values and institutions is the basis of neo-institutionalism (see Brace and Hall 1990; Kato 1996; March and Olsen 1976). Understanding the dynamic nature of policy from a neo-institutional perspective is a valuable method understanding how policy emerges and evolves.

The question arises as to the proper scope of discussion. Should discussion focus solely on the national level or should the state and local level be included? The dynamics of national level institutions are very important in establishing the general movements in many areas of public policy. For that reason, primary focus will be on national-level actors in establishing the overarching rules of the game that govern institutions, establish priorities, aggregate and shape preferences, and the rationality on which preferences are based. State- and local-level institutions will also be discussed in a generic manner, particularly in terms of how overarching emergent policy promotes and shapes social and economic incentives, ultimately impacting bottom up policy innovations. As North (1981) so effectively pointed out, one of the most curious developments in economic history has been the relationship between government institutions, emergent policy incentives and constraints, and economic choice.

THE PRESIDENT

The powers of the president in shaping public policy have evolved tremendously over time. In terms of enumerated powers, the Constitution provides very little formal power for the president. With a few important exceptions, the nineteenth-century presidents were literalists—employing only the limited formal powers granted them in the Constitution.

In the years following the Great Depression, the inherent and implied powers of the presidency were both highlighted and strengthened. A period of intense national crisis, the president was increasingly viewed as responsible for the overall national economic and social health. Through legislative proposals, executive orders, and bureaucratic institutional reform Franklin D. Roosevelt (FDR) built a set of rules and institutions (see Milkis 1993; Reeves 1973) that shaped and continues to shape energy policy and choices within energy markets. While consumed by individuals, the energy supply and access issues became subjects of national policy.

The New Deal–era model of shaping energy policy markets and choices through public policy had long-term consequences. The institutional rules and constraints that emerged from the time period were built on certain assumptions regarding energy supply and demand, as well as a narrow view of political and social values related to economic choice. FDR was probably able to accomplish that which he accomplished in energy policy because of the nature of times, institutional turmoil, and the scope of crises he confronted. Emerging value shifts in the 1950s and 1960s combined with the 1973 oil embargo and subsequent petroleum price shocks opened the floodgates of debate—no longer was the debate simply about economic rehabilitation. Institutional rules, constraints, and assumptions regarding choice were subject to intense scrutiny and the presidency was caught in the middle of the storm. At the time of the 1973 oil embargo, President Nixon faced economic challenges, a rising conflict related to Watergate, and growing malaise with his Vietnam policies. Intellectually led by Milton Friedman, a generation of contemporary free-market scholars called for deregulation of energy policy and allowed for the expanding energy markets. From an opposing direction, ecologically minded individuals and groups saw an opportunity to challenge the New Deal energy paradigm in terms of its environmental impacts, advocating public policy rules and institutions to cultivate the development of renewable green energy sources. From a neo-institutional perspective, there was limited possibility of establishing any lasting set of institutional rules and constraints to shape individual and collective choice: it was a period of deep questioning regarding processes and outcomes.

Facing institutional constraints, Nixon's approach did not deviate significantly from the New Deal model. By executive order, Nixon created the Federal Energy Office (later Federal Energy Administration) to manage the energy crisis and to review energy prices. Nixon also created the Environmental Resources and Development Administration, which was tasked with research and development work with the intention of establishing greater energy independence (Jones and Strahan 1985: 186–87). Possibly for pragmatic reasons rather than a result of idealism, Nixon also presided over significant environmental institutional developments (Flippen 2000), that would, over time, change the rules of the game regarding traditional New Deal–era energy policy. Facing another energy crisis a few years later, Jimmy Carter chose another path—similar in its faith in institutional response to crisis, yet different to the extent that it served as the basis for public-private sector development of a new energy paradigm. Building on the rising environmental movement, Carter created a new set of institutional rules tying environmental policy to energy policy while simultaneously establishing the basis of a new alternative and renewable energy paradigm.

The Reagan administration moved energy policy in the direction of the aforementioned Friedmanesque approach and its emphasis on market-driven solutions. Reagan-era policy was given some credence by energy consumers to the extent that energy prices (particularly energy in the form of petroleum) fell during his presidency. While his successor George H. W. Bush's administration witnessed price increases, prices declined again during the presidency of Bush's successor—Bill Clinton.

Despite low prices, President Clinton's budgets for renewable energy policy were quietly increased, but no significant overt effort was made during the Clinton presidency to move the nation toward a new energy economy. It could be argued that this was consistent with Clinton's Third Way approach to politics and policy; a grand policy strategy was not seen as the appropriate method of solving policy problems. In many ways consistent with North's economic thought, Clinton envisioned at least two general roles for national government in energy policy. First, Clinton thought that national government should define the general purpose and values related to energy policy. Energy policy should be consistent with environmental policy goals—one policy priority should not supercede other policy goals. Clinton also felt that energy policy should promote national security and economic interests (see Clinton 1999: 1743). Second, Clinton saw national government as a facilitator and partner in the innovation processes—in fact, government at all levels should show a commitment to the priorities outlined by elected government. Top-down solutions were eschewed; state and local governments, often in partnership with the private sector and citizen stakeholders, developed policy solutions tailored to needs (Clinton 1995: 1379).

Clinton's energy policy could be viewed as a missed opportunity. Third Way politics does tend to be a more hands-off approach, but it does imply strong leadership commitments. Critics of Clinton's energy policy endeavors come from both sides of the political spectrum. Environmental groups were particularly critical of the petroleum glut and were displeased with the lack of significant increase in research and development monies allocated for alternative and renewable energy. It has been noted that with the exception of the Energy Policy Act of 1992, President Clinton's efforts in the energy policy arena were not of great significance (Jaskow 2002: 105).

From a neo-institutionalism perspective, Clinton's energy policy efforts could be viewed in at least two ways. First, Clinton was a self-described Third Way politician. Third Way thought implies a particular view of government institutions and institutional actors' roles in projected institutional power. As noted earlier, Third Way would imply less projection of power in the policy innovation process and a larger facilitator role. Second, Clinton's presidency was a unique opportunity to study interinstitutional relationships. In 1995, the congressional majorities were Republican for the first time in forty years. The values of the congressional majority and their view of institutional power, rules, and constraints changed. Energy policy was not a foremost issue in the Republicans' new majority—budget and tax issues were of great importance. In terms of energy policy, deregulation of electrical energy utilities and the restructuring of natural gas markets were of great interest to both Congress's new majority as well as to President Clinton.

By the time George W. Bush assumed the presidency in January 2001, crude petroleum prices were steadily increasing due to global demand and reduced inventories led to tightened petroleum markets. Adopting an energy policy strategy not dissimilar from the Reagan presidency, Bush proposed during the 2000 presidential election and during the early months of his presidency, a policy of opening the Arctic National Wildlife Area to petroleum and natural gas exploration (Bernton 2001). The policy was not widely accepted and faced opposition in the new Congress, particularly in the Senate, which as a result of the Senator Jim Jeffords (I-VT) party affiliation change from Republican to Independent, was controlled by Democratic leadership. In the summer of 2001, the National Energy Policy (NEP) report was published and the subsequent National Energy Policy Act was proposed, which would have expanded the nation's energy supply through the aforementioned Arctic exploration and through the development of new generation nuclear power, new generation fuel development (e.g., biodiesel), as well as a variety of forms of green energy.

In the post–September 11, 2001 policy environment, the president used national security as a method of framing the energy policy debate, but the NEP was not supported. In the 2002 midterm election, the Republicans regained control of the U.S. Senate but were unable to successfully pass an energy bill in the 108th Congress. During that same time period, natural gas prices began to increase and there was increased pressure to remedy the situation either through greater imports and/or through energy policy reform (see Allen 2003). During the 2004 presidential election, policy issues debated focused to a significant degree on foreign affairs issues, particularly the U.S. military presence in Iraq and Afghanistan.

In 2005, the Energy Policy Act (EPAct) was passed—the first major piece of energy legislation passed since EPAct of 1992. EPAct 2005 was greeted with little fanfare despite a continued commitment to energy policy reform. In an effort to advance his energy policy agenda, the president visited the National Renewable Energy Laboratory (NREL) in Golden, Colorado, praising NREL's continued work in next generation energy. Earlier in 2005, the Bush administration had recommended budget cuts for NREL—a recommendation that was reversed shortly before his visit to the Colorado site (Curtin 2006). With public opinion polls—an important foundation of presidential institutional power—showing declining support for the president in 2005 and 2006, neo-institutionalism would tend to indicate the president's efforts at major policy reform—energy or otherwise—are constrained (see Daniel 2005; Yeager 2006).

Policy innovation on the part of presidents may result in a fairly large-scale redirection of policy priorities and may have significant economic, social, and political implications. Large-scale reform is not typical. Based on neo-institutional theory, one should not expect large-scale reform from presidents unless there is a crisis situation and strong popular support for policy initiatives. New Deal energy policies, the 1960s and 1970s energy reforms, and the current energy policy proposals led to major changes in the way energy policy and energy markets are structured. New Deal reforms were largely responsible for many of the energy infrastructure developments that remain part of our energy paradigm today. The 1960s and 1970s reforms by presidents Nixon, Ford, and most particularly Carter, serve as the foundation of green energy policy, further emphasized in EPActs 1992 and 2005.

Economically, energy policy innovations often shape the existence of and profitability of markets, establishing a set of general rules, incentives, and priorities. Reagan-era reforms favored the fossil and nuclear energy model over the renewable energy model and declining oil prices made that approach acceptable to the public. Deregulation of energy infrastructure and markets occurred during periods of strong economic growth and low energy prices. Presidents' values and the institutional constraints under which they make choices shape the energy policy choices made. While the Clinton administration environmental policy favored green energy models, institutional constraints combined with Third Way politics may have actually constrained his energy policy efforts. The current Bush administration is likely shaped by the president's values regarding energy. Importantly, the nature of the presidency as an institution often limits large-scale and long-term movement of energy policy in new directions. In short, public opinion impacts policy choices of institutional actors (see Stimson et al. 1995). The president as an energy policy leader is most likely to be in evidence when crises associated with energy policy are widespread and public support for the president as a policy leader is clear and strong.

THE CONGRESS

From a neo-institutional perspective, Congress as an institution operates under certain constraints in terms of institutional rules, processes, and norms. The House of Representatives is subject to a general election every two years; alternatively, the Senate is a continuously legislative body with one-third of its members facing reelection (or defeat) every two years. Thus, members' terms of office have a tremendous impact on their vision and the level of detail and amount of time that can be devoted to any particular policy area. Constituencies differ with senators representing entire states and often representing more national interests. House members operate within districts and their attention is often drawn to parochial constituent interests (see Fenno 1978). Fewer in number, senators are called to serve on a number of policy-related committees and are given greater opportunity to focus attention on the interconnectivity of policy areas. House members often become specialized within a narrow band of policy areas.

In terms of energy policy, institutional structure reduces the possibility that a large percentage of house members will become directly involved in the policy area through their committee membership, although they might see a role for their narrowly focused policy areas to shape energy policy in a manner that best benefits their personal policy interests (e.g., Agriculture Committee members will obviously want to have energy policy subsidies for farmers growing energy-related crops such as rape seed and corn) rather than focusing on the bigger picture. The institutional structure of Congress will reduce the likelihood that the institution will be capable of producing a balanced and coherent national energy policy.

Unlike the presidency, Congress has significant enumerated powers. Enumerated powers are not by themselves the method by which Congress could justify an active involvement in energy policy. Historically, Congress has "found" implied, resulting, and inherent powers that serve as the justification for a variety of policy innovations.[1] During the New Deal era, Congress pursued a national interest, often legitimated through its "necessary and proper" clause power (Article 1, Section 8, Clause 18) and through its powers to regulate commerce (Article 1, Section 8, Clause 3). With Court acquiescence in such landmark cases as *West Coast Hotel v. Parish* (300 U.S. 379 1937), the Court found a constitutional principle via the Fourteenth Amendment that allowed for government to act in a mediating role in shaping private contract relationships. As a result, congressional powers broadened to establish and fund wide-ranging policy priorities.

Value shifts also played a tremendous role in the direction of energy policy. New Deal–era energy policy tended to focus attention on the distributive aspects of national energy policy. Although clearly some states and districts benefited from capital projects in the creation of hydroelectric dams and agriculture water projects, to a great extent these choices were based on geographical considerations. Rural electrification was a policy of national scope and clearly distributive. Institutional changes in Congress (Hinckley 1976) and the rise of green politics paralleled the rise of alternative and renewable energy policy. Environmental and energy industry lobbying played a role in legislator decision making and was quite likely of greater significance in shaping legislator choice than political party affiliation.

As an institution, Congress has and continues to evolve. The 1990s was dominated by a pragmatic president and, in January 1995, a new and relatively well-disciplined Republican majority. With fossil energy prices declining and no general sense of impending energy crises looming, Congress was not appreciably active in shaping energy policy choices. Congressional Republicans criticized President Clinton for releasing oil from the Strategic Petroleum Reserves during the fall and winter of 2000 (see Ivanovich 2000). At the time, the main policy initiative in response came not from Congress, but from a soon-to-be presidential candidate Republican George W. Bush (*San*

Francisco Chronicle 2000). Bush recommended pressuring the Organization of Petroleum Exporting Countries to increase supply and to simultaneously open the Arctic National Wildlife Refuge for oil and natural gas exploration. Regulatory reform advanced by congressional leadership in the fall of 2005 has also shown Congress to be an active institutional actor in shaping the energy policy agenda. In early October 2005, the House of Representatives passed a restructuring of regulatory policy related to petroleum refineries by a slim margin. The bill allows energy producers to build new refinery facilities and streamlines the process by which permitting will occur. Critics claim that the bill also relaxes environmental regulatory related to fuel mixtures and manufacturing. What is particularly interesting is that this bill originated in Congress and was not advanced by the White House, which demonstrates the changing role of Congress as an institution operating within an evolving energy paradigm. Nationally, congressional majorities are effectively standing their ground as part of the existing fossil fuel energy paradigm, adjusting for changing conditions in terms of petroleum quality and supply. Congress is also responding to the interests of their constituencies, who demand reasonably priced liquid fuels for their petroleum-based vehicles.

THE COURTS

The institutional power of the federal courts regarding energy policy is not explicit in Article III. The role of energy and the intra- and interstate transportation of fuels and electricity in the United States is, for the most part, a late nineteenth- and early twentieth-century phenomenon. Early energy policy-related cases were argued within the confines of the judiciary's narrowly defined enumerated powers. In the late nineteenth century, the court system remained timid in terms of taking energy and material related cases. Post-*Lochner* (1905), the judiciary showed itself more willing to enter into disputes between state government and citizens. Although the case had nothing to do with energy policy, the Supreme Court—by taking the case—opened the doors to the expansion of one of its enumerated Article III powers in a way that more directly scrunitized policymaking at the state and local levels and de facto expanded the notion that federal court decisions were linked to the national government's supremacy. The 1937 case *West Coast Hotel v. Parrish* confirmed the judiciary's interest in scrutinizing public policy at all levels of government. Constitutionally, legal theory was now open to an enlarged analysis of the interchange between national regulatory powers and policymaking authority and state and local powers. In essence, the Court more fully disclosed acceptance of judicial positivism in method and decision making rather than a strict constructionist approach.

Over the years, the courts have had a significant role in energy policy. The Court has been particularly interested in regulation of safety with regard to energy policy, as safety issues are central to the public good aspect of energy. While not directly related to energy policy, *New Jersey Steam Navigation Company v. Merchants' Bank of Boston* 47 U.S. 344 (1848) does illustrate the Court's particular and early interest in issues related to energy safety. The case involved the destruction by fire of a steam-operated commercial boat. In the end, the Court sided with the plaintiffs and the decision of a lower court to award damages. The case ultimately turned on the issue of fuel safety as the boat was fitted for wood-burning energy production, but was burning a much hotter "modern" fuel for boats of the time—anthracite coal. Although other deficiencies had been noted in terms of safety equipment that ultimately tied to the issue of the federal court's "admirality jurisdiction," the case provides early evidence that the Court saw a role for itself in reviewing aspects of energy safety, particularly in terms of transportation safety—albeit tangentially and quite possibly with little emphasis beyond the nature of the case.

The Court, however, tightened its level of scrutiny in term of energy-related safety issues in *Champlin Refining Co. v. Corporation Commission of Oklahoma et al.* 286 U.S. 210 (1932). In this case, one of the earliest cases involving the regulation of safety issues related to petroleum refinement, the Court dismissed broadly defined environmental restrictions on the extraction and refining of petroleum. In essence, the Court demonstrated that an early state-level effort to protect the environment from the impact of oil drilling and processing could only occur if statutes were written narrowly and were essentially based on scientific principles related to environmental safety. One could argue that by taking the case and deciding it, the Court opened further the door to national regulation of environmental policy as is most directly related to the issue of energy resource development, processing, and distribution.

The 1970s, a decade in which the petroleum-based energy paradigm experienced a major shock, saw the Court dealing with two prominent cases related to energy safety issues. In *Vermont Yankee Nuclear Power Corp. v. Natural Resources Defense Council, Inc., et alia* 435 U.S. 519 (1978), the Court dealt with questions related to "the proper scope of judicial review of the Atomic Energy Commission's procedures with regards to the licensing of nuclear power plants." In lower court decisions, the commission's rule-making procedures related to nuclear energy fuel management and safety issues were overturned through court decision. In essence, this would have opened the door to further court scrutiny of the nuclear energy process in terms of safety. In a unanimous decision, the late William Rehnquist wrote that the Court of Appeals has improperly developed its own conception of safe reactor process and remanded the case to a lower court to scrutinize the commission's regulatory clarity. The case is significant because it effectively maintained nuclear energy policy as viable as long as rule making and regulatory processes governing this form of alternative energy were rationally constructed and complete. The Court looked to administrative solutions to any lack of clarity or completeness first but was fairly definitive in removing the judicial system from the process or filling in areas of vagueness or rewriting significant portions of regulation and process-related nuclear energy policy.

In the same year, the Court decided the so-called trans-Alaska pipeline rate cases. The Court sought to clarify rate change policies related to the shipment of crude oil and natural gas. In essence, the Court solidified the authority of Interstate Commerce Commission (ICC) in its efforts to manage the pipeline. The commission's ability to adjust rates for rational economic reasons and to require pipeline operators to refund excess rate charges to customers was recognized by the Court. The pipeline cases were critical to the legitimacy of the commission's authority over the transportation of petroleum from Alaska. In a broader sense, the Court established precedence of the ICC to regulate petroleum transportation. Appellate court decision has further solidified its position on pipeline rates in *BP West Coast Products, LLC v. Federal Energy Regulatory Commission* 376 F. 3d 1223 (2004). The Court was careful to balance this decision in relation to the states' power to regulate intrastate energy policy issues.

In *Exxon Corp. et alia v. Governor of Maryland et alia* 437 U.S. 117 (1978), the Court recognized the power of state government to regulate gasoline markets within its borders. The Court found that neither the interstate commerce clause nor the due process clause of the Fourteenth Amendment were violated by Maryland's regulations on petroleum producers' ability to establish gas stations and policy efforts to ensure equity within the gasoline market across various corporate concerns operating fueling station in-state. In essence, the Court established a balance between the interests of the national government in regulating energy transportation and use and the interests of the state in advancing goals not inconsistent with national constitutional interpretation and national policy priorities.

Current energy policy cases have centered on five major areas of concern:

1. the California energy crisis;
2. the Energy Policy Act of 1992;

3. Yucca Mountain nuclear waste depository;
4. the role of science in shaping rule making and rule-making procedures and application, in general; and,
5. the constitutionality of executive branch policy groups.

In terms of the California energy crisis, the Court faced important decisions regarding the supply of energy to residents in California; in essence, the cases challenged the legitimacy of private markets in distributing power and challenged the market equilibrium model by arguing that post facto the corporate energy market should refund profits gained during a period of high demand and reduced supply. The Ninth Circuit ordered FERC (the U.S. Federal Energy Regulatory Commission) to review energy prices and consider refund options in the case *Bonneville Power Administration v. FERC* 422 F. 3d 908 (2005).

While EPAct of 1992 will essentially fade in prominence as EPAct of 2005, the reauthorization of the EPAct will be further explicated in the years ahead. Of particular importance within EPAct of 1992, however, were provisions advancing the creation of market-based energy exchanges independent of power producers. In essence, this was a major first step in the direction that has culminated in EPAct of 2005 and in the dissolution of the Public Utility Commission Holding Company Act of 1935. At the state level, market-based approaches were embraced in California during the Gray Davis administration. In 2000 and 2001, the deregulation plan was severely tested by excess energy demand and an inability to meet that demand—excess demand led to rising prices. California Power Exchange signed many energy contracts with energy providers for above-market prices in an attempt to meet the state's energy demand during particularly warm summer months. In the years following the constrained energy supply phenomenon, California and other energy producers have sought to be released from their above market energy contracts. In *BPA et alia v. Federal Energy Regulatory Commission* 422 F. 3d 908 (2004), the Ninth Circuit Court of Appeals affirmed FERC's statutory authority to hear cases related to potential energy price gouging and governmental agency demands for refunds from power companies. The courts are responding to the privatization of energy policy with a skeptical eye—at least to the degree that equitable market exchanges must occur for privatization to operate efficiently and effectively.

The Court has shown its support for state-level regulatory efforts to define market conditions. In *Puerto Rico Department of Consumer Affairs et alia v. Isla Petroleum Corporation et alia* 485 U.S. 495 (1988), the Supreme Court found that the Commonwealth of Puerto Rico did preempt federal authority in regulating the island's gasoline market and profit margins. In *Oxygenated*

Fuels Association v. Gray Davis, 331 F. 3d 665 (2003), the courts again sided with the states over the federal government. The court found that California's ban on methyl tertiary butyl ether—used to oxygenate fuel, but with potentially harmful impact on the air and groundwater—was not beyond the scope of the state's energy policy authority.

If the Court has made movements toward state-centered federalism, the issue of Yucca Mountain nuclear waste repository provides some clues to the limitations. In cases involving a compelling national interest, the Court will favor the national government over state government concerns. The state of Nevada has been fairly aggressive in its efforts to prevent spent nuclear fuel from being shipped to Yucca Mountain as a final repository for the national nuclear waste. In *Nuclear Energy Institute, Inc. v. Environmental Protection Agency* 62 U.S. App. D.C. 204 (2004), the federal court rejected Nevada's appeal. While the court took exception to the methodology related to the Environmental Protection Agency's ten thousand-year "compliance period," it rejected Nevada's contention that the methodology for site selection was fundamentally flawed.

The federal courts have also recently scrutinized federal rule-making and rule enforcement policies in the following representative cases:

* *Village of Bethany, Illinois, et alia v. Federal Energy Regulatory Commission* 276 F. 3d 934 (2002)
  The Court found in favor of FERC's method of regulating natural gas rate-making. The court argued that while citizens of Bethany, Illinois, were "captive customers," they could ask for renegotiated prices during periods of product price decline; they have the opportunity to negotiate when the provider's next rate-making case occurs.
* *Grover v. U.S. 73* Fed. Apprx. 401 (2003)
  The Court affirmed the oil shale mining moratorium imposed in 1991 and in the EPAct of 1992. Restricting a claimant's ability to access and sell shale oil did not constitute a takings.
* *Shell Petroleum, Inc. v. U.S.* 319 F. 3d 1334 (2003)
  The Court effectively defined tar sands (oil sands) as not being crude oil; therefore, the court affirmed that Shell Petroleum, Inc. was not entitled to crude oil-related tax refunds provided under the Crude Oil Windfall Profits Tax Act of 1980.
* *B & J Oil and Gas v. Federal Energy Regulatory Commission* 359 U.S. App. D.C. 214 (2004)
  The Court found that FERC's allowance for a natural gas firm to expand its operations onto the petitioner's property was acceptable, but that it was acceptable because FERC was able to provide compelling evidence that the

commission's decision is the least invasive method of violating property rights and that the decision was based in objective information and rational decision making.

- *Commonwealth of Massachusetts et alia v. Environmental Protection Agency* 415 F. 3d 50 (2005)
  The EPA has the authority to decline to use rule-making power to regulate "greenhouse gases from motor vehicles" as long as the decision to regulate or not to regulate was based on information on the "frontiers of scientific knowledge."
- *PSI Energy, Inc. & Cincinnati Gas & Electric Co. v. U.S.* 411 F. 3d 1347 (2005)
  A purchaser did not have to pay the enrichment-related processing fees to the U.S. Department of Energy if they sold the material without having used it.

As might be noted, the judiciary largely focuses on issues related to existing large-scale commercial energy systems. Early cases tended to focus on the rules governing private sector interaction, whereas twentieth- and twenty-first-century cases increasingly involve the regulatory power of government in relation to private-sector actors and the management of the energy market.

The lack of a clear linkage between energy policy and the environment is evident in the cases above. However, the courts have shown an interest in the relationship between energy policy and a very broad understanding of social and economic justice. Grossman (2003) made this point fairly well in his delineation of the "three-pronged" test the Court has established to weigh environmental justice cases related to energy policy, outlined in *Friends of the Earth v. Laidlaw Envtl. Services (TOC)* 528 U.S. 167 (2000):

> The Court reiterated that "to satisfy Article III's standing requirements, a plaintiff must show (1) it has suffered an injury in fact that is (a) concrete and particularized and (b) actual or imminent, not conjectural or hypothetical; (2) the injury is fairly traceable to the challenged action of the defendant; and (3) it is likely, as opposed to merely speculative, that the injury will be redressed by a favorable decision."

Grossman (2003) argued that establishing standing can be accomplished, particularly in regions that have experienced significant negative impacts resulting from global climate change. Writing before Hurricanes Katrina and Rita, Grossman anticipated that coastal regions will most likely be able to establish that consequences of global warming have negatively impacted local social and economic conditions. In essence, with standing established, Grossman argued that the current carbon-based energy paradigm can be challenged in terms not just related to economic equity as has been the case in recent years, but on the basic of social equity in terms of the global public health and other externalities that have reduced the potential for sustainable communities of the future.

In neo-institutional terms, the Court has changed and continues to change the "rules of the game" (Buchanan 1991: 20) that govern and/or constrain the decision-making processes with its own branch of government as well as that of other branches of government existing at the national, state, and local levels. In terms of members' values, a movement of the judiciary toward the political Right will likely change the "rules of the game" in a different direction. With the appointment of Chief Justice John Roberts and Associate Justice Samuel Alito Jr., the Court appears firmly dominated by conservative judicial positivists and strict constructionists, which for the moment might make Grossman's point moot. Alternatively, a Left-leaning judiciary might produce quite different results.

STATE- AND LOCAL-LEVEL DEVELOPMENTS

Energy policy at the state and local levels has largely been a function of changing values as well as incentive structures often initiated through national level policy. A string of national-level policy has been largely responsible for opening the door to state- and local-level developments. The latter point is not meant to diminish the bottom-up policy innovations occurring at the state and local levels. State- and local-level innovations are initiated for reasons related to national-level policy, but their uniqueness is indicative of the diversity and entrepreneurial nature of states and local governments.

What motivates state and local governments to develop innovative energy policies? One answer is provided by the politics of federalism. In 1956, Charles Tiebout wrote a now-classic article on tax costs at the local level of government. He argued that the nature of American democracy, particularly federalism, meant that citizens would maintain an important check on tax rates through their economic choices. If tax costs became too excessive in one region or locality, citizens would use their individual economic calculus to make the decision to either stay put or move to a location with a lower tax cost.

Although it is doubtful that Tiebout (1956) anticipated the energy policy questions of the twenty-first century, his argument about tax costs could be extended into other areas, namely the economic feasibility of living in local communities or states based on an energy paradigm that is destined, in the long term, to become very expensive. An important part of sustainable communities is the ability of citizens to be able to afford to live healthy, productive, and meaningful lives. Siek (2002) does a good job outlining the state and local efforts to make sustainability possible and to ultimately "people" communities with individuals seeking an affordable quality of life. Through the use of land use regulation, states and local communities are reinventing themselves. Promoting alternative energy is only one part of these efforts. Other aspects of sustainable communities of the future involves reducing commute times by integrating work and residential zoning in a manner that promotes a quality of life for all. Out-of-home businesses are being encouraged as long as the domestic-based enterprises do not become much larger than cottage industries. Energy conservation through building ordinances is also a critical aspect of state and local regulatory efforts to promote sustainable communities (see Satterthwaite 2002: 1668–69).

The distributive aspects of state and local efforts were discussed earlier, nevertheless, it is important to reflect on the nature of state and local incentives. In essence, the aforementioned regulatory changes are examples of changing the rules of the game—that is, the rules that govern political, social, and economic institutions. The earlier discussion of economic incentives related to the use of alternative energy are examples of government efforts to change the behavior of individuals operating within these changed institutions. Regulation often serves as a "stick," but the sting of regulation can be softened through the use of incentives or "carrots." The movement toward sustainable communities, as discussed in the academic literature, is an attempt to use both cooperative policy efforts to promote a collective good: overreliance on regulation produces outcomes that citizens may adopt but not willingly. Overreliance on incentives alone, however, may reduce the ability to coordinate outcomes in a way that promotes coordination, equity, and efficiency.

INTEREST GROUP INFLUENCES
(PUBLIC INTEREST GROUPS, BUSINESS GROUPS, AND GOVERNMENT GROUPS)

Interest groups come in a variety of different forms and with a multitude of policy preferences. Some groups claim to pursue the public good, seeking to provide generalized benefits for members and nonmembers alike. Other groups seek to gain particularized benefit for themselves as an individual group or for a group of associated organizations. The former groups are known as *public interest groups*; the latter groups generally fall under the general title of *economic groups*. Public interest groups became quite prominent in the 1950s and 1960s and have continued to grow in number and strength ever since. Economic groups are quite prominent, too, and should not be discounted simply because of the celerity with which public interest groups have appeared in society. Economic groups can be single business enterprises, such as Ford Motors, which despite its economic troubles in recent years, remains an important part of the American economic landscape and an influential economic interest. In some cases, economic interests represent a particular economic sector, such as National Association of Realtors. Agriculture economic interests often organize around the product that growers produce—the Cattleman's Association (beef) and Sunkist (citrus).

In 1965, the late Mancur Olson—a famous economist—predicted that economic interest groups were likely to thrive while public interest groups were likely to fail (Olson 1965). The basis of his prediction was related to the nature of the benefits produced by interests. Since public interest groups provided benefits to members and nonmembers alike, there was no strong incentive to join a public interest group—one could simply be a *free rider* and receive the benefit without any financial commitment to the group. Conversely, economic groups provided selective benefits to members *only*. Time has shown that Olson's prediction, while well reasoned, was inaccurate. In part, this is a function of a changing social value base in the United States, as discussed previously. Postmaterialists have become a predominant part of U.S. society and politics (see Inglehart 1990). Increasingly, individuals will make choices to join or support a public interest based on ideological or philosophical reasons rather than simply selective economic benefit. Just as individual values and political behavior have changed, so, too, have the rules of the game governing the U.S. political and public policy landscape. Interest group politics has filled a vacuum left by a weakening political party system in the United States (Pomper 1977: 40).

Interest groups have shown themselves to be a highly effective at influencing public policy at all stages of the policy process. Elected officials, usually members of the two major political parties, often face significant time and resource constraints (and limited incentives) in shaping public policies following the creation of statutes. The budget and committee oversight are tools Congress uses to shape policy poststatute, but time limitations and disincentives often mean that Congress eschews large-scale regular oversight. The president, too, has significant time constraints and relies heavily on appointed officials to represent his views, which has varying impacts on policy poststatute. Interest groups, however, have significant time to follow individual policy arenas and to advance their goals throughout the policy process.

Environmental groups have and will likely remain very influential in energy policy. A centuries-old movement, interest group influence grew tremendously in the 1960s, 1970s, and 1980s in the United States. Environmental groups generally do not seek personal economic benefit from their efforts to protect the environment but tend to be driven by a notion of societal benefit. Many groups have significant resources needed to keep group to promote legislative action. Through lobbying efforts and information campaigns, interest groups shape policy outcomes (see Wright 1990).

Economic groups have played a significant role in shaping the energy policy debate and in a variety of different ways. Rural agrarian counties in the midwestern United States, for instance, have faced economic and social decline for several decades. In the 1970s and 1980s—in part due to rising energy costs—family farmers in the heartland were faced with serious economic depravation. Many of these family farmers were forced to sell their farms, often to large corporate farming interests. Social changes led many midwestern youth to migrate to the cities for greater opportunity, which meant that the next generation of farmers and ranchers disappeared from the rural landscape. Government groups at the state and local levels, such as associations of counties, began to pressure state and national policymakers to promote the use of corn in the alternative energy paradigm—federal ethanol subsidies have played a major role in making this aspect of farming much more profitable and, as a consequence, making farming a more lucrative enterprise.

Urban government interest groups, such as the League of Cities, have also played a major role in shaping regulations and distributive policy incentives to promote sustainable communities. The supply of abundant and cheap energy is the cornerstone of the U.S. city of the twentieth century; curtailing demand but maintaining quality of life will be the challenge of the twenty-first-century U.S. city. In order to accomplish this significant goal, however, government interest groups seek the economic aid of governments at all levels. Given Tiebout's (1956) overarching thesis, it is natural that government interests will jockey for financial opportunities to promote the policy innovations unique to their locale and the needs of their communities in relation to other urban areas.

CHAPTER SUMMARY

Thinking about institutions, policy actors, and institutional differences and communalities for the moment, one faces the realization that the promotion of large-scale change in the energy paradigm is not going to be—in a manner of speaking—a piece of cake. Different institutions have different institutional rules and different incentives. Terms of office, perhaps one of the simplest aspects of institutional differences within and between the legislative and executive branch, for instance, almost immediately leads one to the awareness that the time horizon governing institutional policy goals and preferences are different. The federal courts, of course, have lifetime-appointed judges and justices, which establishes an entirely unique set of institutional rules and incentives (and potentially, disincentives). In a more specific sense, however, the rules of institutions and institutional actor preferences prove to be even more interesting because one is provided with a sense that communalities exist within the initial observation of distinct differences. The economic, social, and political preferences of citizens tend to transcend and overwhelm institutional differences, particularly in times of crisis. If, as some have argued, the time of energy crisis is on us, institutional differences within the policy area will become less apparent as energy costs rise and begin to constrain the micro and macro economic realities governing everyday existence and the ability of the nation to promote a notion of justice that remains a centerpiece of the political, economic, and social milieu for nearly a century.

NOTE

1. As early as 1787, Congress passed the Northwest Ordinance Act, which established a federal presence in promoting K–12 education. Congress also became actively involved in establishing a series of roads and waterways to promote commerce. Neither of these policy examples are clearly enumerated in the Constitution.

WORKS CITED

Allen, E. 2003. Natural Gas Prices Welling Up—Energy Secretary Calls for an Emergency Meeting on the Issue. *San Antonio Express News*, June 11, 1E.

Bernton, H. 2001. Big Battle Brewing over Oil Exploration—New President to Push Drilling in Alaskan Arctic. *The Seattle Times*, January 7, A1.

Brace, P. and Hall, M. 1990. Neo-Institutionalism and Dissent in State Supreme Courts. *Journal of Politics* 52(1): 54–70.

Buchanan, J. 1991. *The Economics and the Ethics of Constitutional Order*. Ann Arbor: University of Michigan.

Clinton, W. 1995. Message to Congress Transmitting the Energy Policy Report. *Weekly Compilation of Presidential Documents*, August 7, 31(31): 1379.

Clinton, W. 1999. Memorandum on the Working Group on International Energy. *Weekly Compilation of Presidential Documents*, September 20, 35(37): 1743.

Curtin, D. 2006. Bush Throws Support behind NREL's Pursuit of Renewable Energy. The President, the First in 28 Years to Visit the Golden Lab, Says He Wants the U.S. to Become Less Dependent on Oil. *The Denver Post*, February 22, A1.

Daniel, C. 2005. Payback Time. *Financial Times*, July 5, Comment & Analysis, 15.

Fenno, R. 1978. *Homestyle: House Members in Their Districts*. Boston: Little, Brown.

Flippen, J. 2000. *Nixon and the Environment*. Albuquerque: University of New Mexico Press.

Grossman, D. 2003. Warming Up to a Not-So-Radical Idea: Tort Based Climate Change Litigation. *Columbia Journal of Environmental Law* 28(1): 2–61.

Hinckley, B. 1976. Seniority 1975: Old Theories Confront New Facts. *British Journal of Political Science* 6(4): 383–99.

Inglehart, R. 1990. *Culture Shift in Advanced Industrial Democracies*. Princeton, NJ: Princeton University Press.

Ivanovich, D. 2000. GOP Criticizes Plan to Tap Reserve. It Won't Raise Heating Oil Supplies, Lawmakers Tell Energy Secretary. *The Houston Chronicle*, September 27, Business Section, 2.

Jaskow, P. 2002. United States Energy Policy during the 1990s. *Current History* 101(653): 105–25.

Jones, C. and Strahan, R. 1985. The Effect of Energy Politics on Congressional and Executive Organization in the 1970s. *Legislative Studies Quarterly* 10(2): 151–79.

Kato, J. 1996. Institutions and Rationality in Politics: Three Varieties of Neo-Institutionalists. *British Journal of Political Science* 26(4): 553–82.

Kingdon, J. 1999. *Agendas, Alternatives, and Public Policy*. New York: Longman.

March, J. and Olsen, J. 1976. *Ambiguity and Choice in Organizations*. Bergen, Norway: Universitetsforlaget.

Milkis, S. 1993. *The President and the Parties: The Transformation of the American Party System since the New Deal*. New York: Oxford University Press.

Murkowski, F. 2000. Drilling Won't Make It Less of a Refuge. *The Washington Post*, December 10, B5.

Neustadt, R. 1960. *Presidential Power: The Politics of Leadership*. New York: Wiley.

North, D. 1981. *Structure and Change in Economic History*. New York: Norton.

Olson, M. 1956. *The Logic of Collective Action: Public Goods and the Theory of Groups*. Cambridge, MA: Harvard University Press.

Pomper, G. 1977. The Decline of the Party in American Democracy. *Political Science Quarterly* 92(1): 21–43.

Reeves, W. 1973. PWA and Competition Administration in the New Deal. *The Journal of American History* 60(2): 357–72.

San Francisco Chronicle. 2000. A Looming Energy Crunch. *San Francisco Chronicle*, September 28, A26.

Satterthwaite, D. 2002. Sustainable Cities or Cities that Contribute to Sustainable Development? *Urban Studies* 34(10): 1667–91.

Siek, A. 2002. Smart Cities: A Detailed Look at Land Use Planning Techniques that are Aimed at Promoting Both Energy and Environmental Conservation. *Albany Law Environmental Outlook Journal* 7: 45–66.

Stimson, J., Mackuen, M., and Erikson, R. 1995. Dynamic Representation. *American Political Science Review* 89(3): 543–65.

Tiebout, C. 1956. A Pure Theory of Local Expenditures. *Journal of Political Economy* 64(October): 416–24.

Wright, G. 1990. Contributions, Lobbying, and Committee Voting in the U.S. House of Representatives. *American Political Science Review* 84(2): 417–38.

Yeager, H. 2006. Bush's Ratings Fall as Katrina Tape Adds to Political Woes. *Financial Times*, March 3, The Americas, 2.

COURT CASES

B & J Oil and Gas v. Federal Energy Regulatory Commission 359 U.S. App. D.C. 214 (2004).

BP West Coast Products, Inc. v. Environmental Protection Agency 362 U.S. App. D.C. 438 (2004).

BP West Coast Products, LLC v. Federal Energy Regulatory Commission 376 F. 3d 1223 (2004).

BPA et alia v. Federal Energy Regulatory Commission 422 F. 3d 908 (2004).

Bonneville Power Administration v. FERC 422 F. 3d 908 (2005).

Brown v. Board of Education 347 U.S. 483 (1954).

Champlin Refining Co. v. Corporation Commission of Oklahoma et al. 286 U.S. 210 (1932).

Commonwealth of Massachusetts et alia v. Environmental Protection Agency 415 F. 3d 50 (2005).

Exxon Corp. et alia v. Governor of Maryland et alia 437 U.S. 117 (1978).

Friends of the Earth v. Laidlaw Envtl. Services (TOC) 528 U.S. 167 (2000).

Grover v. U.S. 73 Fed. Apprx. 401 (2003).

Lochner v. People of State of New York 198 U.S. 45 (1905).

New Jersey Steam Navigation Company v. Merchants' Bank of Boston 47 U.S. 344 (1848).

Nuclear Energy Institute, Inc. v. Environmental Protection Agency 62 U.S. App. D.C. 204 (2004).

Oxygenated Fuels Association v. Gray Davis, 331 F. 3d 665 (2003).

PSI Energy, Inc. & Cincinnati Gas & Electric Co. v. U.S. 411 F. 3d 1347 (2005).

Puerto Rico Department of Consumer Affairs et alia v. Isla Petroleum Corporation et alia 485 U.S. 495 (1988).

Shell Petroleum, Inc. v. U.S. 319 F. 3d 1334 (2003).

Vermont Yankee Nuclear Power Corp. v. Natural Resources Defense Council, Inc., et alia 435 U.S. 519 (1978).

Village of Bethany, Illinois, et alia v. Federal Energy Regulatory Commission 276 F. 3d 934 (2002).

West Coast Hotel v. Parish 300 U.S. 379 (1937).

Chapter Eleven

Practical Demonstration of the Economics of Alternative Energy

INTRODUCTION

Thus far, the book has concentrated on outlining the information needed by individuals participating in the development of a sustainable community. Information is important, but tools are also critical. Tools allow us to take the information that has been learned and apply it to a real world application. The best method of preparing for a real-world situation is to use a simulation tool to practice decision-making capacity. One of the great advantages of a simulation tool is that it is often accessible to all participants working, to create an energy model for sustainable communities.

HYBRID OPTIMIZATION MODEL FOR ELECTRIC RENEWABLES (HOMER)

HOMER® is a shareware program first developed by Peter Lilienthal, chief economist, National Renewable Energy Laboratory (NREL) in 1993. In 1997, Tom Lambert, a professional engineer now at Mistaya Engineering, a Canadian firm, created the first Windows™ -driven version of the program. The program is a multi-iterative simulation program that accepts a user-driven energy system architecture. It allows the user to identify a set of energy-producing technologies as well as energy storage capacity. The output of the program provides the user with an extensive list of policy options with price measures and sensitivities.

A general understanding of alternative energy technologies is required before individuals should operate HOMER®. For business-sector participants and energy policy bureaucrats, alternative energy technology is often well understood. Among citizen stakeholders, however, knowledge levels regarding the technical feasibility of alternative energy is often lacking. As is often the case in other technically driven policy areas, individuals with higher levels of education are more likely to understand alternative energy systems. Thinking back to Douglas and Wildavsky (1983), technical feasibility must coincide with political feasibility—or consent—before policy and associated risks are widely viewed as acceptable. It is perhaps for this reason that alternative energy solutions such as nuclear power—albeit 20 percent of the nation's electrical energy is generated by nuclear energy sources—is left out of this analysis, at least in the direct modeling sense. Instead, the model is intended for a direct analysis of renewable energy systems and perhaps mixed energy systems that include fossil fuel energy generation as supplemental. This chapter will look at several HOMER® models to illustrate program adaptability in use as well as for gauging energy prices for particular energy solutions. The analysis is an opportunity for the reader to become an active learner and policy solution participant through use of HOMER®.

COST INPUTS

Having downloaded HOMER® and become familiar with its operations, the reader will notice that unit price inputs are needed to determine the cost of equipment. This section of the chapter will therefore briefly discuss prices for equipment; in some cases, longitudinal data are available that illustrates changing costs.

Solar Panels

In nominal dollar terms, solar panel prices declined modestly during the first three years of the twenty-first century. Between mid-2001 and mid-2004, solar panel prices dropped from $5.93 to $4.96 per peak watt, which is approximately a 16 percent decline in peak watt price. Since mid-2004, however, nominal peak watt price has increased by roughly 6 percent in nominal dollar terms. In inflation adjusted terms, the price change is perhaps better illustrated. The price of solar panels on a peak watt basis has declined rather precipitously between 2000 and 2004. Prices dropped on a per peak watt basis from slightly more than $6.50/W to a little over $5.10/W. In 2005, prices began to gradually climb, which may be a function of a growing demand for solar panels for residential and commercial uses. The demand might be a function of government incentives as well as prices for energy.

Wind Energy

In a recent analysis, the Danish Wind Industry Association (DWIA) demonstrated that the number of variables impacting the price of wind energy systems are too numerous to hold constant for price comparison purposes. Wind speed, site development issues, and technology conspire to make a year-to-year price comparison quite challenging. Therefore, historic analysis in the price of wind energy would be meaningless and will not be conducted here. Due to rising capital costs driven by tremendous technological breakthroughs as well as the difficulties associated with residential application of wind energy systems, wind energy technology has tended to expand into commercial markets. Wind systems have been and are continually being developed for large-scale wind farms for commercial energy applications. Offshore systems are quite obviously beyond the budget constraint of the vast majority of residential users.

DWIA analysis demonstrates that large-scale wind energy development is perhaps the only economically, politically, and technologically feasible solution to wind energy applications. In studying price/supply curves, the deflection point in terms of cost versus supply indicates that price begins to level off at between two and three cents per kWh for systems generating more than one million kWh per year—large wind turbine systems that cost well over $1 million and whose hub heights may exceed thirty meters.

The current price of a typical residential wind energy unit, however, is available and will be used in this model. The Bergey XL1 is a 1kW system that, under normal conditions, can produce 1kW at 11 m/s (25 mph) and a total of 5.2kWh per day. The tower for the Bergey XL1 is approximately thirty feet high.

The Bergey Corporation lists all of the product specifications on its website, but for the purposes of this analysis price is a primary concern. A new Bergey XL1 costs approximately $3,900. Given that time will likely result in either cheaper products or that a used Bergey XL1 will be purchased, the replacement value of a XL1 was estimated at 35 percent less than a new XL1—a fairly conservative estimate. Installation manuals indicate that the process of erecting a Bergey XL1 can be done by laypersons.

Other Equipment: Batteries, Converter, Generator

When the sun is out and the wind is blowing, energy is collected. Surplus energy must be stored either in batteries in stand-alone nongrid connected systems or can be stored on the electrical energy grid, through something known as net metering—in essence, the alternative energy provider contributes energy to the power grid for use by other energy consumers. In the model discussed below, a stand-alone system is being modeled to demonstrate what kind of system would be needed if the energy producer and consumer were operating independently of other energy consumers and producers.·

In this case, batteries are a very important and necessary part of the energy system. Battery technology is continually improving and will likely become cheaper in the future. In this system, batteries are charged by the electricity produced by the wind and solar energy systems and then discharged whenever demand or load is placed on the system. The batteries will regularly be charged and discharged in the alternative energy model, which places a very heavy strain on the batteries.

For alternative energy systems such as the one being modeled, it is recommended that the system operator use deep-cycle or marine batteries that are designed to withstand continual charging and discharging. In this case, the HOMER® model is using Trojan L16P lead-acid batteries. Information on battery prices is available at the battery manufacturer website for which HOMER® provides direct link access. A number of batteries could be chosen—Trojan batteries happen to be the manufacturer used in sample analyses. Battery prices can be impacted by large orders. If, for example, a community was built on an alternative energy model, thousands of batteries would be needed. Manufacturers often discount the price per unit for large orders and will likely buy used batteries for the lead salvage value and for the battery casings, which can also be recycled. In essence, the battery prices in life-cycle cost analysis would likely significantly decline in a sustainable community model built around alternative energy systems.

In nominal dollars, prices for batteries used in alternative energy system applications have remained fairly stable in the last three years. However, in inflation-adjusted terms, batteries have declined by approximately 6.2 percent since 2002. The price decline is significant, ultimately impacting the affordability of alternative energy systems. The price curve declined somewhat more slowly in 2005, but tended to follow the continued downward trend. It is possible that the decline in price is due to the fact that battery manufacturer capacity has not yet reached a point where increased input costs will be needed and that manufacturer's fixed costs are simply being spread over a larger consumer base. It may also be the case that with increased regular demand for deep-cycle batteries, there are actually reductions in production cost due to improved capacity to recycle battery lead and casings, which result in production cost savings that are passed along to battery consumers. In any event, the price declines are certainly an encouraging sign for battery users for alternative energy systems.

Converters

A converter is a series of power circuits that change the electrical energy into a useable form. Solar panels produce *direct current* (DC), whereas a diesel generator system produces *alternating current* (AC). Therefore, the alternative energy system will likely require AC current; therefore, a converter—a system that converts DC to AC and vice versa—is needed.

MODEL OF ENERGY SYSTEM[1]

Perhaps the most important thing to consider in the model is the primary load. As is discovered in the use of the HOMER® system, the demand for electrical energy in most residences is not constant. Most people leave their homes for work during the day and turn off the lights and all sorts of other electrical gadgets. Heating is turned down because the house does not need to be kept at an especially warm or cool temperature while residents are away. Additionally, less energy is needed at night.

Every community has slightly different load demands. In part, this is due to climatological variations. Hotter summers or cooler winters, for instance, will impact the use of air-conditioning and heating systems. Communities in the United States that are further north rely on more electric light during the winter months than cities and communities in the southern part of the nation where the days are longer. Load demands, therefore, vary in terms of location, climate, and time of day. The HOMER® program allows the individual user to input data on load demand depending upon all of the aforementioned factors. The trick for the individual user is to find the average load data on an hourly basis by month. HOMER® needs to have a clear set of parameters on something known as *peak load*—that period of time when demand on electricity is the highest. Peak load will determine, ultimately, what components are needed in the electrical energy generation system and when they will be needed to meet electrical energy demand.[2]

HOMER® allows the user to select alternative energy off grid components using a simple point-and-click modeling approach. Add/remove buttons allows users to model different types of energy generation for different uses. In some cases, the primary load—the load for home electricity use can be completely eliminated and all energy generated could be used to produce hydrogen gas. The hydrogen could be used to power alternative energy vehicles. Additionally, it is possible to generate electrical energy and to store it on the grid or to at least access the grid for electrical energy as opposed to using a diesel generator as a back-up system. In this case, the model "pretends" that the grid is not available and the system is a stand-alone electrical energy system. The system also allows for thermal as opposed to purely electrical loads. Thermal solar energy, for instance, is particularly valuable during the colder months of the year, but has a year-round application for hot water heaters.

Economic inputs for equipment pricing, replacement costs, initial start-up costs, operations and maintenance costs, emissions, and interest rates are also relevant input variables. Equipment prices are based on the previously conducted analysis in this chapter. Replacement costs, however, are a function of price trend estimates conducted by NREL and other organizations. Although solar prices have fluctuated in the last few years, estimates are the price of solar panels will decrease by as much as 50 percent by 2020. Wind power systems are not expected to decline over the same time period. Therefore, replacement costs are assumed, in this model, to be the same as current costs. For modeling purposes, battery replacement is assumed to be a function of the number of batteries used in a system.

Larger systems will likely benefit from battery lead recovery rebates as well as battery casing rebates. Smaller systems will likely not benefit as greatly. The model assumes that systems employing as many as twenty-four batteries will have a slightly lower (25 percent) replacement cost than one battery unit system. The data on converter price trends are not clear; therefore, the model assumes no variation in unit replacement costs. Many of these costs are included in sample analyses that are part of the HOMER® program and in the sample programs available with this text, but will be updated based on changing economic and price conditions. For example, interest rate sensitivities are changing as interest rates on money change over time. At this point, the model is using 5 and 7 percent interest rate sensitivities. The user can also adjust prices and economic sensitivities, so the program is truly user friendly and responsive to user input.

Load factors in this analysis assume that the homeowner is making *no change* in his or her residential electrical energy uses. In essence, the homeowner is simply converting his or her home to alternative energy to meet an assumed load. To make alternative energy affordable, homeowners will likely want to reduce electrical load demands through a variety of techniques. First, a homeowner could purchase a solar refrigerator that operates on solar energy. Second, it is feasible to use a more efficient hot water heater. Third, a homeowner could retrofit an existing air-conditioning system to improve efficiency. Fourth, improved insulation and energy-efficient windows could be used to reduce heating and cooling costs in the winter and summer months. Finally, energy-efficient light bulbs might reduce energy consumption. It should be noted that these efficiencies are currently available options.

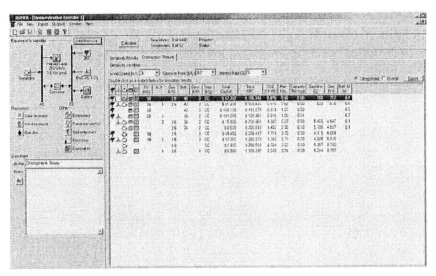

Figure 11.1. HOMER® Optimum Systems

Once all of the inputs and exponents are selected, then clicking "calculate" will start a multi-iterative modeling process. The diagram in figure 11.1 illustrates a screen shot of the actual HOMER® output. The analysis took nearly nine hours and completed over eight hundred thousand simulations before arriving at the optimum solutions based on equipment inclusion, price sensitivities, wind speed, fuel prices, battery lifetime minimums, and interest rates. The screen shot represents average wind speeds for the location, current fuel prices, and approximates current interest rates using approximate current yield rates on ten-year treasury bond (actual: 5.55 percent; see www.bloomberg .com/, accessed March 29, 2006).

The best-case scenario relies on photovoltaics (PVs). Obviously, a wind system would require greater room to install and is the second-best scenario. Ground mounting would be needed and sufficient space—assuming a backyard installation—to install guide wires. Additionally, a small and highly concentrated neighborhood might frown on wind turbine installation for at least two reasons. First, there might be noise associated with the operation of a wind turbine. Second, a wind turbine mounted nine or ten meters above the ground might pose a serious liability to the owner should the turbine and tower collapse and injure an individual or damage or destroy property.

The optimum system would also require ten 1kW solar panels. The PV panels could be installed on rooftops and given modern tile-like solar panel installations, the PV system would be unobtrusive and nearly invisible to other homeowners in the neighborhood. Solar panels, however, require direct sunlight and almost zero shading. Without getting into too many of the details of electrical wiring, suffice it to say that solar PVs are wired in series, which means that shading could render a solar panel completely useless. One of the interesting features of many homes, however, is the presence of shade trees, which reduce residential dependency on air conditioning during the summer months; however, the shade tree is also the potential "enemy" of the effective solar panel. Simple observation of this dilemma is certainly worth remembering when planning out the energy independent home or community and requiring long-term planning regarding neighborhood and home aesthetics as well as energy generation capacity via renewable or alternative means.

Since the optimum system is utilizing two DC solar electrical energy generation systems, it is necessary to install a converter to create the AC current required for most residential appliances. The system requires a 3kW converter. As this system operates a 2.6kW generator on average less than one hour per day, there is a substantial demand for battery storage capacity. In this model, 40 Trojan L16P batteries are used which individually have a capacity of 390 amp hours (Ah). Wired in series, the batteries have a capacity of 4,680 Ah. The dimensions of the batteries are: L 10 3/8" x W 7 1/8" x H 11 3/16". Space would have to be made available for battery storage, serviceability, and ventilation. According to the HOMER® model, the batteries will be drawn on most heavily during the fall months.

It should be noted that the load data for HOMER® in the following analyses are not modeling for winter heating, which is assumed to be in the form of either oil, wood, or natural gas-fired central-heating units. As discussed previously, heat pumps could serve to reduce the energy demand for thermal energy during the winter months. The model does, however, model for air-conditioning demands since it takes into account electrical energy load factors for residential homes—in this instance, in the state of Delaware.

The model demonstrates that a conservative estimate of energy production stands at 29.1kWh of electrical energy produced to serve a peak residential load of 3.59kWh. This means that if the homeowner has a grid connection, it is possible that 1,284kWh or 23 percent of the produced electricity could become surplus electricity for other purposes. In building the sustainable community, it is possible that surplus generation could be used to operate clean industries within the community or elsewhere. Emissions data from the optimum system demonstrate that there may not be a direct emissions benefit from the optimum energy system. According to the emissions report, 1,764 kg/yr of carbon dioxide emissions will result from the optimum system.[3] To eliminate this problem of emissions, the second-best system listed, which has ten 1 kW solar PV panels and one wind energy system could be chosen. The system costs approximately $4,000 more in terms of net present costs, but it does reduce greenhouse gas emissions. There is an important trade-off to consider from both the individual energy producer/consumer level as well as from a sustainable community level: namely, what is the purpose of utilizing alternative/renewable energy systems? Clean air is a public good that could be promoted through alternative/renewable zero emission energy production and consumption practices.

The purpose of the HOMER® demonstration above is to increase awareness of a policy tool.[4] Sensitivity analysis could be conducted using the program to demonstrate net present costs of an energy model as well as the cost of electricity. For example, the best-case scenario in the case presented produces electricity at $0.67/kWh, which is substantially more expensive than electricity purchased from the average power company.

CHAPTER SUMMARY

The purpose of this chapter is to demonstrate the utility of HOMER® and its accessibility to the average consumer. Sustainable community development, which promotes collectively derived solutions to public goods dilemmas, such as clean air, public health, and economic sustainability and justice, requires that all citizens have some access to the decision-making process. Additionally, citizens and professional policymakers need to work together to develop a sense of the trade-offs associated with the adoption and development of new generation energy generation and consumption patterns. In this case, HOMER® analysis is demonstrates how one might go about comparing alternatives and getting a sense of the true costs of these alternatives in the present day as well as over time. Although not the subject of this chapter, there is also a need to reduce load demand through the adoption of currently available technologies such as high efficiency air-conditioning and heating systems, insulation systems, heat pumps, solar refrigerators, and low wattage lighting systems. Reducing load demand will reduce the cost of new generation electrical energy systems and storage units or batteries.

NOTES

1. Special thanks to Peter Lilienthal, Ph.D., at the National Renewable Energy Laboratory, and Tom Lambert, P.E., M.Sc., Mistaya Engineering for reviewing my HOMER® outputs.

2. For the purposes of this study, hourly data were gathered from Delaware Electric.
3. The estimated annual electricity production for the optimum system is 16,433 kWh/yr. It produces 0.11 kg/kWh (.24 lbs/kWh). According to a published report, average conventional electricity used in a residence produces 1.64 lbs/kWh (see Cadmus Group 1998; EPA 2006; ICF Consulting 1999). The green energy system detailed here produces approximately one-seventh the CO_2 emissions.
4. An accompanying CD will provide more details on how to use HOMER® to generate cost scenarios for specific locations.

WORKS CITED

Cadmus Group. 1998. *Regional Electricity Emissions Factors Final Report*. Boston: Cadmus Group, Inc.

Douglas, Mary and Wildavsky, Aaron. 1983. *Risk and Culture: An Essay on the Selection of Technological and Environmental Dangers*. Berkeley: University of California Press.

Environmental Protection Agency (EPA). 2006. *Personal GHC Calculator Assumptions and References*. yosemite.epa.gov/oar/globalwarming.nsf/content/Resource CenterToolsCalculatorAssumptions.html, accessed May 4, 2006.

ICF Consulting. 1999. *Emissions Factors, Global Warming Potentials, Unit Conversions, Emissions, and Related facts*. www.p2pays.org/ref/07/06861.pdf, accessed May 4, 2006.

WEB SITE

www.bloomberg.com/, accessed March 29, 2006.

Conclusion

The Future of Alternative Energy

INTRODUCTION

As the world braces itself for peak oil prices and rapid increases in energy demands, the common mantra emanating from energy policy circles is "The future of alternative energy is *now*." It is very possible that the future has arrived and that the theoretical developments within the sustainable communities circles must now be implemented. Entering into the foray as either a wild-eyed idealist or as a deeply skeptical cynic is perhaps shortsighted and unproductive. In the former case, it is very likely that for several years and perhaps decades, the idealist will experience frustration and disappointment. Conversely, the cynic will either slow down the process that must occur or, at the very least, be surprised to find that sustainable communities are feasible and will produce, in the long run, substantial quantities of needed energy to a world with growing demands.

The best position to take, perhaps, is that of the cautious optimist, remaining open to the idea that sustainable communities will emerge and will produce positive outcomes that extend beyond simply the issues of energy supply and demand. Caution, of course, also requires that one enter into the transitional phase with one's eyes open, aware of problems that must be addressed in the short run and the benefits that will likely emerge in the long run.

INFORMATION: LIMITATIONS AND PROSPECTS

One of the biggest areas of concern is the issue of safety. While policy experts often understand safety issues related to alternative energy and fuels, the public often relies on decades' old information, much of it highly inaccurate. Although solar and wind energy power systems are generally thought to be safe, alternative fuels such as hydrogen are an explosive issue for the public despite the fact that some believe it is safer than gasoline.

In *Consumer Views on Transportation and Energy*, Elyse Steiner (2003) confirms many of the public concerns with hydrogen, demonstrating a serious technical information gap between policymakers, energy policy administrators, and citizens. In a 1998 public opinion poll conducted for National Renewable Energy Laboratory (NREL), the plurality of citizen survey respondents indicated that electric cars and trucks would replace internal combustion engines (ICEs). Only 3 percent of those surveyed indicated that hydrogen-powered cars and trucks would likely replace fossil fuel vehicles (Steiner 2003: 32). In a 2000 survey commissioned by NREL, approximately 14 percent of survey respondents thought hydrogen was the best fuel to replace ICEs (Steiner 2003: 33), whereas 27 percent of respondents felt it was the "worst fuel" replacement—electricity was still rated as the optimum fuel replacement with over 50 percent of respondents identifying it as the "best fuel." In a 2003 Harris Poll, the primary concern regarding alternative fuels was their demonstrated safety for passengers and drivers. Respondents were also concerned about the costs of vehicles and fuel, as well as the convenience of refueling. Although 67 percent were concerned with fuel emissions, there are other concerns related to cost.

Information accessibility and availability are also problematic in terms of conducting economic and technical feasibility analyses. While the HOMER® simulation presented in the previous chapter operated quite smoothly, there was tremendous technical report analysis conducted prior to the simulation being calculated. Many prices for turn-key systems are not entirely apparent due to proprietary issues. Additionally, assumptions regarding true costs are often unclear due to the presence of short- or long-term subsidies provided by the government.

ALTERNATIVE ENERGY/FUELS AND
THE ECONOMICS OF PUBLIC HEALTH

As discussed in chapter 11, choosing alternative energy systems must not focus simply on the cheapest method of generating electricity. In a sustainable community energy model, hidden costs must also be considered. In the "optimum" electricity generation model, the use of a generator actually produced a significant amount of carbon dioxide. These emissions from fossil fuel-powered generators represent a "cost" that should be factored into energy system choices. Evidence from the public health and environmental justice literature focuses attention on the long-term costs of carbon dioxide (CO_2) emissions.

Public health issues begin at home. Studies now confirm that automobile emissions are found inside homes. Airborne particulate matter, carbon dioxide and monoxide, and sulfur dioxide can move from the garage space to the home's living space. Children traveling to school via school bus are also subjected to the negative effects of petrochemicals. Bus engines usually operate on diesel fuel. Scientists have found substantial evidence linking school bus emissions with increased prevalence of childhood asthma. A dangerous illness of the upper respiratory system, asthma and particularly childhood asthma, has been linked to permanent physical disability and even death. In prolonged exposure studies of tollbooth operators, emissions from automobiles have been linked to reduced mental functioning. Still other studies have shown that fossil fuel emissions are linked to significant damage to croplands and to buildings. Emissions reduce crop yields and cause substantial plant losses because of air and water pollution originating from fossil fuel emissions. Forest lands and native ecosystems are damaged by fossil fuel use. Buildings are heavily damaged by the use of fossil fuels. Carbon emissions often form a thick sooty coating on buildings. Combined with rain water or humidity, sulfur dioxide is corrosive to concrete and steel. Still in use in some countries, lead-based gasoline produces harmful airborne lead emissions.

Public health economics has produced some reports on the economic costs associated with carbon based fuel usage. In 1997, N. Eyre and colleagues developed a method for determining economic impacts of fossil fuel use in transportation (see figure C.1). What is particularly interesting about the Eyre et al. study is that it takes a "bottom up" approach to analyzing fossil fuel impacts (1997: 5).

ALTERNATIVE ENERGY/FUELS AND
ECONOMIC DEVELOPMENT

Alternative energy is an important part of any future domestic or international economic development. The postindustrial world is the greatest user of petroleum and other forms of fossil energy to sustain quality of life. The service industry economic base in the postindustrial world is heavily dependent on fossil energy to move people to destination resorts and to provide consumers with goods and services. Quality-of-life issues drove citizens—with the requisite economic means—to the suburbs in the postindustrial world. Lifestyles became, by necessity, more mobile. The decline of the traditional

Damage Costs of Transport Emissions

Emission	Impact	Damage Costs (in p/km)					
		Rural Emissions			Urban Emissions		
		Petrol	Gas	Diesel	Petrol	Gas	Diesel
Carbon dioxide	Global warming	0.093	0.073	0.068	0.109	0.085	0.095
Methane	Global warming	0.000	0.005	0.000	0.000	0.006	0.000
Nitrous oxide	Global warming	0.003	0.003	0.001	0.006	0.006	0.001
Carbon monoxide	Global warming	0.001	0.001	0.000	0.003	0.001	0.001
Particulates	Health	0.003	0.000	0.151	0.003	0.000	1.692
Particulates	Building materials	0.000	0.000	0.003	0.000	0.000	0.035
Sulphur dioxide	Health	0.024	0.001	0.014	0.173	0.001	0.182
Sulphur dioxide	Crops	0.000	0.000	0.000	0.000	0.000	0.000
Sulphur dioxide	Timber	0.018	0.001	0.011	0.021	0.001	0.015
Sulphur dioxide	Building materials	0.005	0.000	0.003	0.036	0.000	0.038
Sulphur airosol	Health	0.033	0.001	0.020	0.038	0.001	0.027
Oxides of nitrogen	Health	0.013	0.007	0.029	0.076	0.054	0.113
Oxides of nitrogen	Timber	0.022	0.013	0.051	0.036	0.023	0.049
Oxides of nitrogen	Building materials	0.006	0.003	0.013	0.034	0.024	0.051
Nitrate aeroll	Health	0.101	0.057	0.228	0.163	0.103	0.219
Ozone from NO_2	Health	0.045	0.026	0.102	0.073	0.046	0.098
Ozone from NO_2	Crops	0.003	0.001	0.006	0.004	0.003	0.006
Benzene	Health	0.012	0.000	0.004	0.126	0.001	0.052
Ozone from VOC	Health	0.110	0.017	0.017	0.145	0.018	0.041
Ozone from VOC	Crops	0.006	0.001	0.001	0.008	0.001	0.002
Nonmethane VOC	Global warming	0.003	0.000	0.000	0.003	0.000	0.001
Sub-totals		0.500	0.211	0.723	1.060	0.375	2.717

Figure C.1.   Social and Environmental Costs of Fossil Energy Use
Source: From Eyre et al. (1997: 17).

urban mix of residential and commercial enterprises meant that walking, bik-
ing, and the use of mass transportation would become less relevant for many
citizens. Paradoxically, the postindustrial world is also very critical of the externalities created by their own lifestyle—in short, these individuals want the
freedom of movement, but they tend to eschew the environmental costs associated with their choices. Therefore, there is a growing demand to promote alternative energy policy and alternative fuel policy, particularly since energy
prices have increased in recent years. Economically, alternative energy and
fuel availability will increase fuel availability. Ceteris paribus, increased supply will result in decreased costs. Additionally, alternative energy and fuels
that have lower levels of emissions and are less dangerous to use will be aesthetically pleasing to many citizens in postindustrial society.

The NREL has developed a relatively user-friendly Excel spreadsheet-based calculator to analyze direct and indirect economic benefits from developing alternative energy production systems. Known as JEDI™ (Job and Economic Development Impact), the program is based in economic assumptions from another program called IMPLAN™. JEDI™ has been used to show the direct and indirect economic benefits for wind power generation systems in Montana (see Costani 2004), which projected significant positive economic impact in terms of easement payments, employment development, and tax base growth. The program inputs are location-dependent, which requires the user to do some "homework" in order to get a sense of project costs, local taxes, property values, et cetera. Nevertheless, it is safe to say that economic development benefits can be quite significant for postindustrial as well as newly industrialized nations and communities.

CONCLUSION

In chapter 9, the Douglas and Wildavsky (1983) model was discussed in relation to the acceptability or unacceptability of risk. In the cases of "peak oil" and alternative energy, it is commonly concluded that waiting to get a better a sense of the future price of oil and when and if the peak price has occurred is a risk that is not acceptable—economically, socially, and politically. Solutions, however, carry with them a high level of risk as well. Some alternatives, such as nuclear energy, may be too risky and possibly too expensive—in terms of waste removal and storage—to be long-term solutions to the energy supply issue. Other solutions may be nearly free of environmental degradation and public health impacts, but may be too expensive and may increase the risk that energy supply needs will not be met. For example, the sun goes behind clouds and the wind may not blow at the rate that will be needed to meet energy needs. Energy storage systems such as batteries may work, but these renewable energy sources may require greater capital expenditure than is assumed simply to make certain that all weather-related contingencies are adequately addressed when it comes to energy supply.

The economics of alternative energy have become clearer. Policy analysts are developing a better sense of the costs of fossil energy in terms of public health, agriculture, and infrastructure. Additionally, economic benefits are being analyzed systematically using programs such as JEDI™. Nevertheless, economic coherency in relation to alternative energy systems remains underdeveloped in many areas. Benefits and costs are a function of location, tax costs, and populations. As Nobel Laureate Douglass North noted, the institutional constraints that guide economic choices are dynamic—what might be a wise economic choice today might not be so tomorrow. As another Nobel Laureate, James Buchanan might note, the "rules of game" have changed and have impacted public and private choices.

Currently, many choices related to possibly short-term policy inputs from government. Tax incentives and direct cash payments are two methods by which government has influenced the economics of alternative energy. Capital costs are declining but remain fairly high. Factoring in the economic benefits of reducing fossil fuel economic dependency is one way of demonstrating a benefit/cost ratio that might be more palatable, but it is important to understand the nature of the benefits being produced and the nature of the costs being assumed. In the case of alternative energy tax and direct cash payment incentives, the politics of public policy would indicate that the nature of the direct policy benefit is individualized, which means that either interest group or clientele politics shapes the policy environment. Benefits calculations at the present time, however, tend to place a great deal of emphasis on generalized benefits to the community or nation as a whole—this form of benefit is more correctly associated with what James Q. Wilson (1990) would call majoritarian politics and policy. Majoritarian benefits would not likely be a part of the economic determination of an individual consumer of tax and or direct cash payment benefits. Ceteris paribus, economic choices to move toward certain forms of alternative energy might not be made without government incentives.

Economic coherency, therefore, is critical to making the leap to several of the forms of alternative energy presented here. At the moment, coherency is lacking in terms of the inputs into benefit/cost analyses as well as a lack of consistent understanding of the technological and political feasibility of various forms of energy. With greater coherency and understanding, however, it is more likely that choices made by individuals will be shaped by community needs. Many benefits can be maximized and many costs minimized by focusing on inputs and anticipated outputs beyond the individual level. The alternative energy paradigm might very well maximize our individual benefit but will require, as is the case in so many other policy areas in the modern world, a greater focus on energy as being a form of public good.

WORKS CITED

Costani, M. 2004. *Quantifying the Economic Development Impacts of Wind Power in Six Rural Montana Counties Using NREL's JEDI Model*. Golden, CO: National Renewable Energy Laboratory. NREL/SR-500-36414.

Douglas, Mary and Wildavsky, Aaron. 1983. *Risk and Culture: An Essay on the Selection of Technological and Environmental Dangers.* Berkeley: University of California Press.

Eyre, N. et al. 1997. Fuel Location Effects on the Damage Costs of Transport. *Journal of Transport Economics and Policy* 31(1): 5–24.

Steiner, Elyse. 2003. *Consumer Views on Transportation and Energy.* Golden, CO: National Renewable Energy Laboratory.

Wilson, J. 1990. *Bureaucracy: What Government Agencies Do and Why They Do It.* New York: Basic Books.

Index

Adams, William, 88. *See also* solar
    energy
AeroVironment Corporation, 163
Afghanistan, 193
agenda setting, 27–28
Arctic National Wildlife Refuge
    (ANWR), 15, 27, 56, 78–79, 81, 192,
    196
Arizona, 53, 138
Atomic Energy Act 75, 179. *See also*
    EPAct of 1992; Nuclear energy

baby boomers, 7–8, 11, 14, 69. *See also*
    culture shift
Batteries for Advanced Transportation
    Technologies Program (BATT), 145
Bell Laboratories, 100
Bequerel, Alexandre, 88. *See also* solar
    energy
Bezdek, Roger, xi
biodiesel, *See* clean diesel
biomass, 41, 54–55, 58*n*, 76, 151
Bonneville Environmental Foundation,
    98
Brush, Charles F., 47. *See also* wind
    energy
Bush, George H. W., 14, 63–64, 72–76;
    EPAct of 1992, 14, 63–64, 72–76

Bush, George W., 12, 15, 27, 54, 73,
    78–79, 80–83, 148, 151, 160–161,
    192–96; alternative energy, 78,
    160–61; ANWR, 15, 27, 78, 195–96;
    EPAct of 1992, 73; EPAct of 2005,
    80–83, 193; Freedom Car, 151;
    Hydrogen Initiative, 54, 78, 151;
    NREL, 193; natural gas, 148

California, 66–67, 72, 97, 100*n*, 117–18,
    126, 130, 138, 150–51, 184, 198–99
California Innovation Group, 67
Canada, 39, 74, 80, 133, 162, 164, 204
Carter, Jimmy, 13, 32, 65, 68–69, 191,
    194; Clean Water Act, 13;
    Department of Energy, 13, 32;
    energy crisis, 13, 32, 65, 68–69;
    renewable energy, 191, 194
China, 1, 17, 28, 47, 147
Chernobyl, 180, 182
clean air, 2, 10, 25, 80, 150–51,
    160–61, 183–84, 217. *see also* Clean
    Air Act. *see also* public good
Clean Air Act, 10, 13, 150–51, 160–61,
    183–84
Clean Cities, 14, 73. *See also* EPAct of
    1992
clean diesel, 149–50, 161

Clean Urban Transport for Europe
(CUTE), 161
Clean Water Act, 13
Clinton, William J., 12, 14, 16, 34,
63–64, 73, 76–78, 83, 115, 152,
191–92, 194; alternative energy, 14,
83, 191; emissions standards, 16;
Executive Order 13149, 77; EPAct of
1992, 14, 73, 76–78; Hydrogen
Initiative, 78; reinventing
government, 34, 63–64; Strategic
Petroleum Reserve, 195; sustainable
communities, 152; Third Way
politics, 191–92; Wind Power
America, 115
coal, 1, 4–7, 14, 39, 40–42, 53–55, 67,
73–74, 76, 81, 131–32, 145–47, 149,
165n, 181–84, 197; 1980s, 14; clean
coal, 76, 81; coal-based diesel, 76;
coalbed methane, 149; Coal
Research Center, 173; derived
liquids, 41–42; electricity
generation, 181–84; emissions,
131–32; gasification and gases,
41–42, 53–54, 73–74; geothermal
energy, 67; India, 131; Poland,
131–32; policy, 40–42
Congress, 11–12, 14–15, 27, 30–31, 67,
69, 75–76, 79–80, 179, 192–96, 204,
206n
Council of American Building Officials
(CABO), 73. See also EPAct of 1992
culture shift 7–9. See also green
movement

Danish Wind Energy Association, 211
Denmark, 47

economic development, 94-95
Endangered Species Act, 13
Energy Policy Act of 1992, 14, 63–64,
72–77, 83, 150–51, 194, 199; George
H. W. Bush, 63; Clean Cities, 73;
Clinton, 14, 63, 77; ethanol, 150–51;
nuclear energy, 75

Energy Policy Act of 2005, 15, 78–81,
83, 193, 199–200; alternative energy,
80; clean air, 80; domestic petroleum
and gas, 81
ethanol, 81
FERC, 81; nuclear energy, 81, 83; tax
credits, 80–81
Energy Production and Conservation
Act, 68
Energy Research and Development
Administration, 67–68
Environmental Protection Agency
(EPA), 14–16, 34, 80, 150, 180–81,
200–1; emissions standards, 34,
201, 184; National Ambient Air
Quality Standards, 16; nuclear
waste, 179–81, 200; Tier II
standards, 184; Transportation, 150,
161
Environmental Resources and
Development Administration, 191
Ericsson, John, 88. See also solar
energy
ethanol, 42, 54, 74, 76, 81, 150–51,
160–61, 205
Europe, 8–9, 47, 55–56, 59n, 69, 77,
103, 132, 134, 151–52, 161–64;
biodiesel, 55; Clean Urban Transport
for Europe (CUTE), 161
commitment to alternative energy, 69,
77, 151–52, 161; Eastern Europe,
132, 134; European Wind Energy
Association, 56; green movement,
8–9; hydrogen fuel cells, 161–164;
mixed oxide fuel, 183; public
opinion, 56; wind energy, 47, 103
European Wind Energy Association
(EWEA), 56
Executive Order 13149, *Greening the
Government through Federal Fleet
and Transportation Efficiency*, 77

Federal Energy Office, 191
Federal Energy Regulatory Commission
(FERC), 75, 80–81, 199, 200

Federal Solar Energy Technologies
Program (SETP), 95–97
Foley, Thomas S. Institute for Public
Policy and Public Service, xi
Fossil energy, 1–2, 5–7, 10, 42–43, 95,
98–99, 103, 133–34, 156, 158–60,
164, 171, 184, 186, 195, 221–23;
public good, 2; emissions, 221; global
use, 1–2; Office of Fossil Energy,
160; U.S. supply and use, 5–7
Fuel Cells; alkaline (AFC), 157; direct
methanol (DMFC), 157–58; molten
carbonate (MCFC), 158–59;
phosphoric acid (PAFC), 159–60;
proton exchange membrane
(PEMFC), 152–56; solid oxide
(SOFC), 159–60
Franssen, Herman xi

geothermal energy, xv, 12, 36, 41–42,
51–54, 58–59*n*, 67, 76, 123–40, 139,
153, 155, 176; binary, 52, 125, 131;
direct use, 125, 126, 128–30, 138;
Federal programs, 135–37; flash
steam, 124–25, 128, 130–31, 140*n;*
heat pumps, 125–26; safety, 52
Germany, 47, 77, 183
Geysers Project, 130
goods; public, 1–2, 24–25, 35–36, 139,
197, 203, 217, 224; marketable
private, 24, 36, 65
Gore, Albert Jr., 27, 77
Great Depression, 7, 69, 190
green movement, 8–9. *See also* culture
shift
green tags, 98, 139

Hawaii, 98–99, 139,
heat pumps, 125–126. *See also*
geothermal energy
Herzik, Eric B., xv
HOMER®, xv, xvi*n*, 209–18
hydroelectric dams, 175–77
hydrogen, 15, 57, 78, 81, 83, 151–52.
See also EPAct of 2005

Hydrogen Initiative, 15, 78, 81, 83,
151–52
public opinion, 56–57
Hythane™, 54, 59*n*, 149–50, 161,

Iowa, 117–18
India, 28, 131,
Iraq, 18, 78, 193

Japan, 1, 130
Jones, Charles O., 26

Kansas, 117–18
Kingdon, John, xii, 27–28. *See also*
Agenda Setting

Lambert, Tom, xv, 209, 218*n*
League of Cities, 205
Lilienthal, Peter, xv, 209, 218*n*
Lithuania, 133–34
Lovrich, Nicholas P., xv, 116–17;
democratic policymaking 116–17

Maine, 118, 138
Maine Maritime Academy, 138
Massachusetts, 118, 201, 208
methyl tertiary-butyl ether (MTBE), 151
Michigan, 118, 182
Miller-Warren Energy Lifeline Act, 66,
83
Million Solar Roofs Initiative, 97
Minnesota, 117
Montana, 117–18, 223
Mouchout, Auguste, 87. *See also* solar
energy
Murphy, James, xv

National Academy of Sciences, 76
National Ambient Air Quality Standards
(NAAQS), 16. *See also*
Environmental Protection Agency
National Environmental Policy Act
(NEPA), 13, 15
National Research Council (NRC), 78,
180–81

National Renewable Energy Laboratory
(NREL) xv, xvi*n*, 12, 16–17, 57,
119*n*, 140*n*, 145, 193; creation, 12;
EPAct of 2005, 193; HOMER®, 209;
public opinion studies, 57, 220
National Science Foundation, 67
natural gas, 4–6, 14–15, 28, 39–40, 42,
54, 55, 59, 70–71, 73–74, 145,
147–49, 151, 161, 173, 192, 196,
198, 200, 216; ANWR, 192, 196;
Courts, 198, 200; EPAct of 1992,
73–74; FERC, 148; HOMER®, 216;
Hythane™, 54, 149, 161
National Energy Technology
Laboratory, 173; alternative energy,
147–48; composition, 148;
importation, 148; liquefied, 42, 74,
148; Natural Gas Act of 1938, 148;
Natural Gas Act of 1978, 73–74;
stripping, 59, 156; supply, 28, 54,
148, 193
Nebraska, 117
Nevada, 34, 54, 95, 138, 156, 179–80,
200
New Environmental Paradigm, 10–11
Nixon, Richard, 11, 13, 63, 68, 71, 190,
194; 1972 election, 11; energy crisis,
190; Environmental Resources and
Development Administration, 191;
Federal Energy Office, 191; policy
innovation, 63; policy reform, 13,
191; price controls, 68, 71
North Carolina, 118
North Dakota, 117–18
North, Douglass C., xiii, 58–59,
189–92, 223–24
nuclear energy, xii, 8, 10, 13–14, 26, 39,
41, 55–56, 66–67, 70–72, 75, 81, 83,
178–85; Idaho National Laboratories,
81; Price Anderson Act, 72; public
opinion, 55–56, 67; mixed oxide
fuels, 183; spent nuclear fuel (SNF),
178–79, 183; Three Mile Island,
13–14, 70, 71–72; EPAct of 1992,
75; Ronald Reagan, 66

Nuclear Energy Task Force, 184–85
Nuclear Waste Policy Act, 179

Occupational Safety and Health Act,
13
Office of Energy Efficiency and
Renewable Energy (EERE), 16
oil embargo, 5, 13, 47, 190
Oklahoma, 117, 173, 197
Oregon, 98, 118, 138, 176
Organization of Petroleum Exporting
Countries (OPEC), 5, 13–15, 17,
27–28, 195–96; oil shortage, 13–14,
27–28

passive microwave, 111
Pennsylvania, 41, 44, 46, 58*n*, 186*n*
Petroleum, xi, 3–10, 13, 17–18, 21, 24,
27–28, 42–43, 54–55, 66, 70–83,
103, 145–49, 153, 161, 172–73,
186*n*, 190, 192, 195–96, 197–200,
221; ANWR, 15, 27, 78, 80–81;
alternative fuels, 54–55, 73–75;
courts, 197–200; global demand,
17–18, 21; OPEC, 5, 27–28; peak,
xi, 3; Petroleum Experiment Station,
173; prices, 70–71, 76–77, 103,
146–47, 190, 192; Strategic
Petroleum Reserve, 195–96; Supply,
66, 192, 195–96; Tars, 24, 80–81,
146; U.S. consumption, 5; U.S.
production, 6
photovoltaic cells (PV), 41, 43–46, 75,
87–94, 152, 154–56, 215
Poland, 131–32
policy making, 21–36; bottom-up,
33–34; collaborative, 35; top-down,
33–34
policy process, 25–31
policy types, 31–33
presidency, 11, 13–15, 68, 72, 77, 115,
190–94
Price Anderson Act, 72. *See also*
nuclear energy
price controls, 68, 71, 73, 83*n*

public good, 1, 24–25, 35–36, 197, 103–205, 217. *See also* goods; HOMER® analysis, 217
interest groups, 203–5
defined, 1; distribution of, 24; marketable, 2, 139; tragedy of the commons, 1; fossil energy, 2
public health, 202–3
public opinion, 55–56, 57, 220; Gallup Organization, 55–56
public utility commissions, 65–66, 71
Public Utility Holding Company Act (PUHCA), 75, 80, 83
Public Utility Regulatory Policy Act (PURPA), 42, 69–71, 75, 80, 83

Quantum Corporation, 163

radar altimeters, 111
Reagan, Ronald, 8–10, 14, 30, 63, 66, 70–72, 77, 83*n*, 179, 191–92, 194
Renewable Portfolio Standards, 97, 138
Romer, Brian, xv–xvi
Roosevelt, Franklin D., 69, 82, 114, 190, rural electrification, 33, 47, 104, 175, 195

Sacramento Municipal Utility District, 94–95. *See also* solar energy
Saudi Arabia, xi
scatterometers, 111
September 11, 15, 54–56, 78, 83, 103, 193
silicon, 43, 88–91, 100*n*. *See also* solar energy
Simmons, Matt, xi
Simon, Raffi G., xv
Simon, Susan M., xv
solar energy, 12, 26, 35, 41–44, 55–56, 67, 70–71, 74, 87–100, 118, 137–39, 153–54, 176, 210, 212–14; collaborative policymaking 35; EPAct of 1992, 76; EPAct of 2005, 81; HOMER®, 209; public perceptions, 55–56; refrigerators,

215; thermal, 43–45, 76, 87–89, 91–93, 98, 214
South Dakota, 117
Spokane, Washington, xi–xii
Steel, Brent, xv
sustainable communities, xii–xiii, 35, 36*n*, 58, 94–95, 139, 152, 164, 184, 202–3, 205, 209, 219; collaborative policy making, 35; courts, 202; Federal Solar Energy Technologies Program, 95–96; fuels, 164; geothermal energy, 139; hidden costs, 164; Hydrogen Initiative, 152; League of Cities, 205; nuclear energy, 184; public health, 202–3; solar energy, 94–95. *See also* economic development
synthetic aperature radar (SAR), 111

Teledyne, 163
Texas, 28, 117–18, 138, 148
tragedy of the commons, 1. *See also* public good
Transportation Technology Transfer Center, xv
Three Mile Island, xii, 14, 70, 179–80, 185. *See also* Nuclear energy
Turkey, 111, 126, 128

U.S. Air Force, 163
U.S. Army, 158
U.S. Department of Energy (DOE), 6, 13, 16–17, 32, 34, 45, 52, 55, 74–76, 79–81, 87, 96, 123, 134–36, 145, 150, 158, 178–80, 183–84; 21st Century Truck Program 17; alternative fuels, 55, 74–75, 150; Batteries for Advanced Transportation Technologies (BATT), 145; constituent policy, 32; demonstration projects, 16–17, 75; EERE, 16; Energy Information Agency (EIA), 87, 92, 161; EPAct of 1992, 74–76; EPAct of 2005, 80–81; environmental awareness, 16;

establishment of, 13; Freedom Car,
17; fuel cells, 158; geothermal
resources, 52, 123, 134–36; NREL,
16; national laboratories, 17; natural
gas, 6; nuclear power, 183–84;
nuclear waste, 178–80; Office of
Fossil Energy, 159–60; policy
innovation, 45, 68, 76, 79
U.S. Enrichment Corporation, 75. *See
also* EPAct of 1992; Nuclear energy
U.S. Navy, 68, 179
U.S. Supreme Court, 15, 196, 199
Union of Soviet Socialist Republics
(USSR), 47, 133, 180
Utah, 138

Vestas Corporation, 47, 154. *See also*
Wind

Warren-Alquist State Energy Resources
Conservation and Development Act,
66
Washington State, xi, xii, xv, 98–99,
138, 178–79,

Washington State Supreme Court,
179
Washington State University, xi
wicked problems, 2–3, 7
wind, 12, 22, 26, 35, 41–42, 46–51, 53,
55–56, 67, 70–71, 76, 81, 91, 98,
101–19, 138–39, 153–56, 176,
211–12, 214–17, 220, 223; Danish
Wind Industry Association (DWIA),
211; energy, 12, 41–42, 46–51, 53,
67, 70–71, 91, 103–19, 138–39,
153–56, 176, 211–12, 214, 217, 220,
223; EPAct of 1992, 76; EPAct of
2005, 81; HOMER®, 209–18;
patterns, 111–12; public opinion,
55–56; Residential Solar and Wind
Energy Credit, 98; safety, 48–50;
turbines, 22, 26, 35, 46–71, 104–6,
108–10, 118, 154, 211, 215; Wind
Power America, 114–16
World Bank, 133–34
Wyoming, 117

Yucca Mountain, 179–80, 199–200